VISICALC®
FOR SCIENCE AND ENGINEERING

VISICALC®
FOR SCIENCE AND ENGINEERING

by
Stanley R. Trost
Charles Pomernacki

SYBEX®

Berkeley • Paris • Düsseldorf

Cover art by Patrice Larue
Layout and design by Margaret Cusick

Library of Congress Card Number: 83-60045
ISBN 0-89588-096-2
Printed in the United States of America
10 9 8 7 6 5 4 3 2 1

To Elaine and Niela

Table of Contents

Acknowledgements

The authors wish to acknowledge the contributions to this book made by the following Sybex Staff members: Rudolph Langer and Rodnay Zaks, advice and encouragement; James Compton, editing; Elaine Foster and Donna Scanlon, word processing; Joel Kreisman, technical support; Valerie Brewster, typesetting; Margie Cusick, artwork and design; and Cheryl Vega, Michael Howard, and Lorraine Aochi, proofreading.

David Lichtblau also provided valuable technical assistance.

Introduction

VisiCalc® is one of the most popular and most important programs available for personal computers. VisiCalc has found wide use in business for the following reasons:

1. it is easy to learn,
2. it is fun to use,
3. it can be applied to a wide variety of problems, and
4. its "what if" capability allows rapid investigation of alternative scenarios.

This book will show you how to apply VisiCalc to a wide variety of scientific and engineering problems; you will find that the reasons for VisiCalc's success as a business tool apply equally to its engineering applications.

We have found that many problems can be solved with the VisiCalc program far more rapidly than by writing a program in a high-level language such as BASIC or Fortran. VisiCalc is especially useful for generating design tables; our emphasis in this book is on the design and application of these tables. To generate a design table you will specify input parameters and an incremental value by which a particular parameter will be varied. The VisiCalc program will then generate a design table based on the increments you specify. This method allows you to study the effects of temperature, pressure, frequency, etc., on a particular design.

Before using this book, you should have a working knowledge of the VisiCalc program. This basic information may be obtained in the VisiCalc reference manual or in Douglas Hergert's *Mastering VisiCalc.* Information on VisiCalc applications, layout of forms, and the documentation methods used in this book is available in Stanley R. Trost's *Doing Business With VisiCalc.*

VisiCalc for Science & Engineering will be useful to professional scientists and engineers, students, and hobbyists. Professionals will find the library of ready-to-run design spreadsheets useful in their everyday work. They will also see how to extend the techniques of this book to other applications. Students will learn how to use VisiCalc as a scientific and engineering tool and set up problems for computer solution. Hobbyists will find many of the applications to be interesting, instructive and useful.

Each example in this book begins with a description of the application which follows. In many of the examples, we include the formulas that the VisiCalc instructions implement. We then discuss the variable data that we enter, and the units of measurement. Each example concludes with a sample problem and a series of instructions (comparable to a program listing in a programming language) which will enable you to reproduce the spreadsheet.

The spreadsheet instructions are interpreted as follows:

```
B15: +C5*C6/A12
/R:B16.B25:NNR
```

First, move the cursor to B15: using the cursor controls or the GOTO command. Then type in the formula:

```
+C5*C6/A12
```

You follow the formula by pressing the RETURN or ENTER key. Now, leave the cursor at B15: and use the REPLICATE command by typing:

```
/R:B16.B25:NNR
```

Let's see how we convert a simple formula to VisiCalc notation and create a design table. We will convert degrees Centigrade to Fahrenheit using the formula:

$$°F = 1.8 + °C + 32$$

We can store an initial Centigrade temperature value at cell C5 and an incremental temperature value at C6. The formulas:

```
C8: +C5
C9: +C8+C6
/R:C10.C20:RN
```

provide 13 values of Centigrade temperature in cells C8 to C20. We obtain equivalent Fahrenheit temperatures by writing:

```
D8:1.8 * C8 + 32
/R:D9.D20:R
```

We have run all the examples in this book on an Apple II computer and an IBM PC. The examples will run on all computers with standard VisiCalc. You can easily convert them to run with SuperCalc and other spreadsheet programs.

1 Mathematics and Statistics

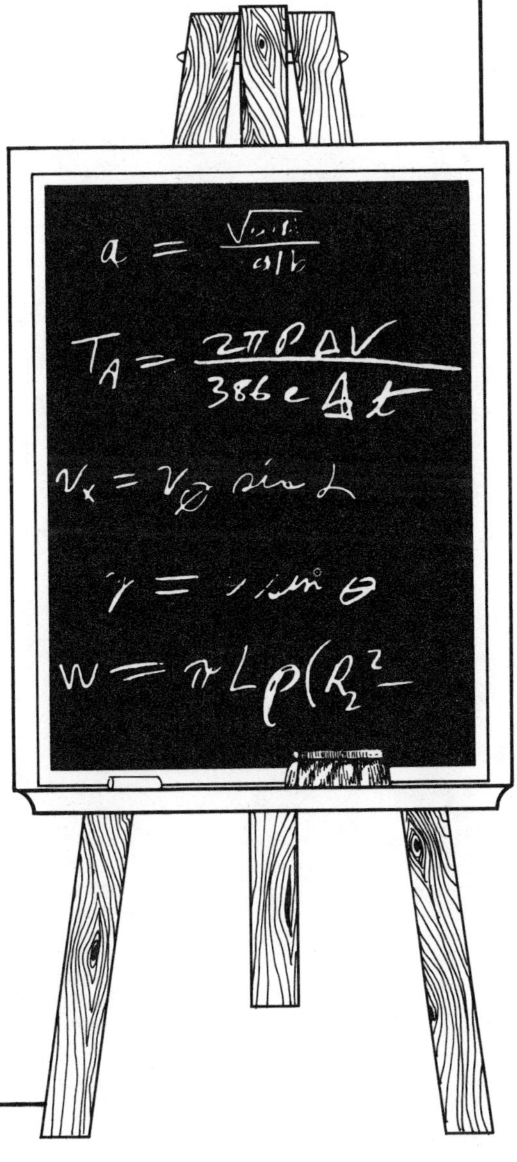

QUADRATIC EQUATION

PURPOSE

This spreadsheet can be used to calculate the solutions to a quadratic equation of the form:

$$Ax^2 + Bx + C = 0$$

The equation has three possible solution sets; this spreadsheet uses the @IF function to test the sign of

$$B^2 - 4AC$$

All three solutions are generated; however, the TRUE symbol on the spreadsheet indicates the correct solution among the three.

DATA ENTRY

You must enter the coefficients A, B, and C. The spreadsheet will then display the parameter Z, which is the value that results from subtracting $4 \times AC$ from B^2. The three solution sets will also be displayed, with the TRUE indicator above the correct solution.

EXAMPLE

You wish to find the solution to the equation

$$2x^2 + 2x + 1 = 0$$

The resulting spreadsheet is displayed in Figure 1.1. Its spreadsheet instructions are shown in Figure 1.2.

```
          A          B          C          D          E          F
 1
 2
 3
 4
 5            QUADRATIC  EQUATION  SOLUTION
 6
 7            A(X^2)+BX+C=0
 8
 9  ENTER COEFFICIENTS
10
11  A=                 2
12  B=                 2
13  C=                 1
14
15  *************************************************
16
17  Z=                -4
18
19                 FALSE       FALSE                TRUE
20  ROOT 1         ERROR       -.5                  -.5 2.000000J
21  ROOT 2         ERROR       -.5                  -.5 -2.00000J
```

Figure 1.1: *Quadratic Equation Spreadsheet*

```
B5:QUADRATIC      C5:" EQUATION     D5:" SOLUTION
B7:A(X^2)+BX      C7:" + C=0
A9:ENTER COE      B9:FFICIENTS
A11:A=
A12:B=
A13:C=
A15:/-*
/R:B15.E15
A17:Z=
   B17:+B12*B12-(4*B11*B13)
B19:@IF(B17>0,@TRUE,@FALSE)
C19:@IF(B17=0,@TRUE,@FALSE)
E19:@IF(B17<0,@TRUE,@FALSE)
A20:ROOT 1
   B20:(-B12+(B17^(.5)))/(2*B11)
   C20:-B12/(2*B11)
```

Figure 1.2: *Spreadsheet Instructions: Quadratic Equation*

```
    E20:-.5*B12/B11
    F20:@ABS(B17)^.5
    G20:J
A21:ROOT 2
    B21:(-B12-(B17^.5))/(2*B11)
    C21:-B12/(2*B11)
    E21:-.5*B12/B11
    F21:@ABS(B17)^.5*(-)
    G21:J
```

Figure 1.2, Continued

DEGREE-TO-RADIAN CONVERSION

PURPOSE

This spreadsheet converts a series of angles measured in degrees into their equivalent angles expressed in radians. The VisiCalc program calculates all trigonometric functions in radian form; thus, this spreadsheet and the following one will be of use in making angular conversions. The spreadsheet displays a table of equivalent angles.

DATA ENTRY

The degree-to-radian spreadsheet has been set up to convert ten angles expressed in degrees into their radian form. You must enter an initial angle and an incremental step size.

EXAMPLE

You wish to convert to their radian form ten angles starting at 0 degrees and increasing in twenty-degree increments. The resulting spreadsheet and its instructions are shown in Figures 1.3 and 1.4.

```
        A          B          C          D          E          F
 1
 2
 3              DEGREE TO RADIAN CONVERSION
 4
 5              ENTER PARAMETERS:
 6
 7              ANGLE            0.00
 8              ANGLE INC       20.00
 9
10
11
12 ***************************************************************
13
14                 DEGREES     RADIAN
15                    0.00       0.00
16                   20.00       0.35
17                   40.00       0.70
18                   60.00       1.05
19                   80.00       1.40
20                  100.00       1.75
21                  120.00       2.09
22                  140.00       2.44
23                  160.00       2.79
24                  180.00       3.14
25                  200.00       3.49
```

Figure 1.3: *Degree-to-Radian Conversion Spreadsheet*

```
/GF$
B3:DEGREE TO     C3:" RADIAN C     D3:ONVERSION
B5:ENTER PAR     C5:AMETERS:
B7:ANGLE
B8:ANGLE INC
A12:/ – *
/R:B12.F12
B14:/FR DEGREES
```

Figure 1.4: *Spreadsheet Instructions: Degree-to-Radian Conversion*

```
    B15: +C7
    B16: +C8+B15
    /R:B17.B25:NR
C14:/FR RADIAN
    C15:@PI/180*B15
    /R:C16.C25:R
```

Figure 1.4, Continued

RADIAN-TO-DEGREE CONVERSION

PURPOSE

This spreadsheet is a companion to the prior example; it will convert a series of angles expressed in radians into their equivalent form in degrees.

DATA ENTRY

You must enter an initial angle in radians and a radian increment; the spreadsheet will then display a series of eleven conversions.

EXAMPLE

You wish to convert a series of angles expressed in radians into equivalent measurements in degrees. You would like the table to start with 0 radians and increase in 0.5-radian increments. The spreadsheet is shown in Figure 1.5; the instructions for the spreadsheet are shown in Figure 1.6.

```
          A          B          C          D          E          F
 1
 2
 3               RADIAN TO DEGREE CONVERSION
 4
 5               ENTER PARAMETERS:
 6
 7               RADIAN          0.00
 8               RAD INC         0.50
 9
10
11
12 *************************************************************
13
14                   RADIAN   DEGREES
15                     0.00      0.00
16                     0.50     28.65
17                     1.00     57.30
18                     1.50     85.94
19                     2.00    114.59
20                     2.50    143.24
21                     3.00    171.89
22                     3.50    200.54
23                     4.00    229.18
24                     4.50    257.83
25                     5.00    286.48
```

Figure 1.5: *Radian-to-Degree Conversion Spreadsheet*

```
/GF$
B3:RADIAN TO       C3:" DEGREE C     D3:ONVERSION
B5:ENTER PAR       C5:AMETERS:
B7:RADIAN
B8:RADIAN IN       C8:C
A12:/ - *
/R:B12.F12
B14:/FR RADIAN
   B15: +C7
   B16: +C8 + B15
```

Figure 1.6: *Spreadsheet Instructions: Radian-to-Degree Conversion*

```
/R:B17.B25:NR
C14:/FR DEGREES
C15:180/@PI*B15
/R:C16.C25:R
```

Figure 1.6, Continued

RECTANGULAR-TO-POLAR COORDINATES

PURPOSE

There is often a need for converting measurements expressed in rectangular coordinates into equivalent measurements in polar coordinates. The spreadsheet in this example will convert a series of rectangular coordinates into their equivalent polar form. The equations used for the coordinate conversion are:

$$R = \sqrt{(x^2 + y^2)}$$

$$\theta = \arctan(y/x)$$

DATA ENTRY

You will enter initial x and y values, and x and y increments into the spreadsheet. The VisiCalc program will then generate a series of coordinate conversions.

EXAMPLE

You are going to prepare a spreadsheet which generates coordinate conversions from rectangular to polar coordinates. The initial x and y coordinates are 1.00. The x coordinate is incremented by one unit. The resulting spreadsheet and its instructions are shown in Figures 1.7 and 1.8.

```
          A           B           C           D           E           F
 1
 2
 3            RECTANGULAR TO POLAR COORDINATES
 4
 5            ENTER PARAMETERS:
 6
 7            X                  1.00
 8            X INC              1.00
 9            Y                  1.00
10            Y INC              0.00
11
12 *************************************************************
13
14                      X           Y           R       THETA
15                   1.00        1.00        1.41       45.00
16                   2.00        1.00        2.24       26.57
17                   3.00        1.00        3.16       18.43
18                   4.00        1.00        4.12       14.04
19                   5.00        1.00        5.10       11.31
20                   6.00        1.00        6.08        9.46
21                   7.00        1.00        7.07        8.13
22                   8.00        1.00        8.06        7.13
23                   9.00        1.00        9.06        6.34
24                  10.00        1.00       10.05        5.71
25                  11.00        1.00       11.05        5.19
```

Figure 1.7: *Rectangular-to-Polar Coordinates Spreadsheet*

```
/GF$
B3:RECTANGUL    C3:AR TO POL    D3:AR COORDI    E3:NATES
B5:ENTER PAR    C5:AMETERS:
B7:X
B8:X INC
B9:Y
B10:Y INC
```

Figure 1.8: *Spreadsheet Instructions: Rectangular-to-Polar Coordinates*

```
A12:/ - *
/R:B12.F12
B14:/FR X
   B15: +C7
   B16: +B15+C8
   /R:B17.B25:RN
C14:/FR Y
   C15: +C9
   C16: +C15+C10
   /R:C16.C25:RN
D14:/FR R
   D15:((B15^2))+((C15^2))^.5
   /R:D16.D25:RR
E14:/FR THETA
   E15:180/@PI*@ATAN(C15/B15)
   /R:E16.E25:RR
```

Figure 1.8, Continued

POLAR-TO-RECTANGULAR COORDINATES

PURPOSE

This spreadsheet is a companion to the previous one; it performs a conversion from polar to rectangular coordinates using the following formulas:

$$x = r \cos \theta$$

$$y = r \sin \theta$$

DATA ENTRY

The spreadsheet is set up to perform a series of ten conversions. You must specify the initial r and theta values and the amounts by which these values are to be incremented.

EXAMPLE

You will set up a spreadsheet which converts coordinates with a constant r value of 5.00 and theta values ranging from zero to 150 degrees in 15-degree increments. The resulting spreadsheet is shown in Figure 1.9; its instructions appear in Figure 1.10.

	A	B	C	D	E	F
1						
2						
3		POLAR TO RECTANGULAR COORDINATES				
4						
5		ENTER PARAMETERS:				
6						
7		R	5.00			
8		R INC	0.00			
9		THETA	0.00			
10		THETA INC	15.00			
11						
12	**					
13						
14		R	THETA	Y	X	
15		5.00	0.00	0.00	5.00	
16		5.00	15.00	1.29	4.83	
17		5.00	30.00	2.50	4.33	
18		5.00	45.00	3.54	3.54	
19		5.00	60.00	4.33	2.50	
20		5.00	75.00	4.83	1.29	
21		5.00	90.00	5.00	0.00	
22		5.00	105.00	4.83	-1.29	
23		5.00	120.00	4.33	-2.50	
24		5.00	135.00	3.54	-3.54	
25		5.00	150.00	2.50	-4.33	

Figure 1.9: *Polar-to-Rectangular Coordinates Spreadsheet*

```
/GF$
B3:POLAR TO      C3:RECTANGUL    D3:AR COORDI    E3:NATES
B5:ENTER PAR     C5:AMETERS:
B7:R
B8:R INC
B9:THETA
B10:THETA INC
A12:/ - *
/R:B12.F12
B14:/FR R
   B15: + C7
   B16: + B15 + C8
   /R:B17.B25:RN
C14:/FR THETA
   C15: + C9
   C16: + C15 + C10
   /R:C17.C25:RN
D14:/FR Y
   D15: + B15 * @SIN(@PI/180 * C15)
   /R:D16.D25:RR
E14:/FR X
E15: + B15 * @COS(@PI/180 * C15)
/R:E16.E25:RR
```

Figure 1.10: *Spreadsheet Instructions: Polar-to-Rectangular Coordinates*

HYPERBOLIC FUNCTIONS

PURPOSE

The VisiCalc program does not have hyperbolic functions built into it. Some applications of the program may require the use of these functions; this spreadsheet will show you how to compute the hyperbolic functions and at the same time will generate a table of these functions.

DATA ENTRY

We have set up a spreadsheet to generate the sinh, cosh, and tanh for ten values of *x*. You must specify an initial *x* and an *x* increment.

	A	B	C	D	E	F
1						
2						
3		HYPERBOLIC FUNCTIONS				
4						
5		ENTER PARAMETERS:				
6						
7		X	0.00			
8		X INC	0.20			
9						
10						
11						
12	**********	**********	**********	**********	**********	**********
13						
14		X	SINH X	COSH X	TANH X	
15		0.00	0.00	1.00	0.00	
16		0.20	0.20	1.02	0.20	
17		0.40	0.41	1.08	0.38	
18		0.60	0.64	1.19	0.54	
19		0.80	0.89	1.34	0.66	
20		1.00	1.18	1.54	0.76	
21		1.20	1.51	1.81	0.83	
22		1.40	1.90	2.15	0.89	
23		1.60	2.38	2.58	0.92	
24		1.80	2.94	3.11	0.95	
25		2.00	3.63	3.76	0.96	

Figure 1.11: *Hyperbolic Functions Spreadsheet*

EXAMPLE

Figure 1.11 (on the opposite page) shows a table of hyperbolic functions beginning with $x = 0.00$ and increasing in 0.20 increments. The instructions for this spreadsheet are displayed in Figure 1.12.

```
/GF$
B3:HYPERBOLI     C3:C FUNCTIO     D3:NS
B5:ENTER PAR     C5:AMETERS:
B7:X
B8:X INC
A12:/ - *
/R:B12.F12
B14:/FR X
   B15: +C7
   B16: +B15+C8
   /R:B17.B25:RN
C14:/FR SINH X
   C15:.5*(@EXP(B15) - @EXP( - B15))
   /R:C16.C25:RR
D14:/FR COSH X
   D15:.5*(@EXP(B15) + @EXP( - B15))
   /R:D16.D25:RR
E14:/FR TANH X
   E15: +C15/D15
   /R:E16.E25:RR
```

Figure 1.12: *Spreadsheet Instructions: Hyperbolic Functions*

STANDARD DEVIATION

PURPOSE

We will design a spreadsheet which will calculate the standard deviation for a series of measurements. This spreadsheet has been set up for 15 measurements; however, this may be easily extended to any number that you desire. We will also calculate the average value of the data, the maximum and minimum data points, and the number of measurements that are entered.

```
            A          B          C          D          E          F
 1
 2
 3          STANDARD DEVIATION
 4
 5          ENTER X  VALUES
 6
 7 ***********************************************************************
 8
 9                   NUM.          X   (X-XAV)(X-XAV)^2
10                   1.00      87.10     -0.51 .2635111
11                   2.00      87.80      0.19 .0348444
12                   3.00      87.50     -0.11 .0128444
13                   4.00      87.40     -0.21 .0455111
14                   5.00      87.40     -0.21 .0455111
15                   6.00      86.90     -0.71 .5088444
16                   7.00      88.40      0.79 .6188444
17                   8.00      88.10      0.49 .2368444
18                   9.00      87.90      0.29 .0821778
19                  10.00      87.80      0.19 .0348444
20                  11.00      88.30      0.69 .4715111
21                  12.00      88.20      0.59 .3441778
22                  13.00      86.80     -0.81 .6615111
23                  14.00      87.20     -0.41 .1708444
24                  15.00      87.40     -0.21 .0455111
25
26                                      SUM SQ 3.577333
27          N                  15.00       MAX    88.40
28          AVE                87.61       MIN    86.80
29                                     STD DEV .4883532
```

Figure 1.13: *Standard Deviation Spreadsheet*

DATA ENTRY

You must enter up to 15 data values in the *x* column. The required quantities will then be computed automatically.

EXAMPLE

We have entered 15 *x* values in the spreadsheet example of Figure 1.13 (shown on the opposite page). The instructions for the spreadsheet are shown in Figure 1.14.

```
/GF$
B3:STANDARD        C3:DEVIATION
B5:ENTER X         C5:VALUES
A7:/ – *
/R:B7.F7
B9:/FR NUM
   B10:1
   B11:1 + B10
   /R:B12.B24:R
C9:/FR X
D9:/FR " (X – XAV)
   D10: + C10 – C28
   /R:D11.D24:RN
E9:/FR "(X – XAV) ^2
   E10:/FG  + D10 * D10
   /R:E11.E24:RR
D26:/FR SUM SQ
   E26:/FG @SUM(E10.E24)
B27:N
   C27:@COUNT(C10.C24)
B28:AVE
   C28:@AVERAGE(C10.C24)
D27:/FR MAX
```

Figure 1.14: *Spreadsheet Instructions: Standard Deviation*

```
      E27:@MAX(C10.C24)
   D28:/FR MIN
      E28:@MIN(C10.C24)
   D29:/FR STD DEV
      E29:/FG (E26/C27) ^.5
```

Figure 1.4, Continued

CHI SQUARE TEST

PURPOSE

The Chi Square test is useful in a number of data analysis applications. It provides a measure of the correlation between observed and expected values in experimental measurements. The spreadsheet in this example is set up to calculate the Chi Square value for a series of data. The spreadsheet may be easily expanded to accommodate more data.

DATA ENTRY

You must enter the observed or measured data together with the expected data values. The spreadsheet will then generate the Chi Square statistic, the number of data values, and the maximum and minimum values.

EXAMPLE

A sample spreadsheet is presented in Figure 1.15. The spreadsheet instructions are shown in Figure 1.16.

```
      A         B         C         D         E         F
1
2
3               CHI SQUARE TEST
4
5               ENTER OBSERVED AND EXPECTED VALUES
6
7   ***********************************************************
8
9                         O         E(O-E)^2/E
10                      115       120 .2083333
11                      375       360 .6250000
12                      460       480 .8333333
13                      250       240 .4166667
14                      245       229 1.117904
15                      690       688 .0058140
16                      920       918 .0043573
17                      440       459 .7864924
18                      125       131 .2748092
19                      375       393 .8244275
20                      540       524 .4885496
21                      270       262 .2442748
22                       15        19 .8421053
23                       60        58 .0689655
24                       80        78 .0512821
25                       40        39 .0256410
26
27                          CHI 2     6.817956
28                          NUM             16
29                          MAX            920
30                          MIN             15
```

Figure 1.15: *Chi Square Test Spreadsheet*

```
B3:CHI SQUAR      C3:E TEST
B5:ENTER OBS      C5:ERVED AND
D5:" EXPECTED     E5:" VALUES
A7:/ – *
/R:B7.F7
B9:/FR O
C9:/FR E
D9:/FR "(O – E) ^2/E
   D10:(B10 – C10) ^2/C10
   /R:D11.D25:RRR
```

Figure 1.16: *Spreadsheet Instructions: Chi Square Test*

```
C27:CHI 2
   D27:@SUM(D10.D25)
C28:NUM
   D28:@COUNT(B10.B25)
C29:MAX
   D29:@MAX(B10.B25)
C30:MIN
   D30:@MIN(B10.B25)
```

Figure 1.6, Continued

THREE-POINT MOVING AVERAGE

PURPOSE

A three-point moving average is a data smoothing and forecasting method which is often used for the filtering of data. This spreadsheet will compute both an unweighted and a weighted three-point moving average for a series of data points. The unweighted average gives equal weight to each of the three prior data points; the weighted method gives a weight of three to the most recent data point, two to the second most recent point, and one to the third most recent point.

DATA ENTRY

You will enter the historical data which is to be smoothed. The form has been set up to evaluate twelve data points; you may extend the form to accommodate more data points.

EXAMPLE

A sample spreadsheet with cyclical data is shown in Figure 1.17; the instructions for this spreadsheet appear in Figure 1.18.

```
          A         B         C         D         E         F
 1
 2
 3              THREE POINT MOVING AVERAGE
 4
 5              ENTER DATA
 6
 7  ***********************************************************
 8
 9                   DATA     UNWTD       WTD
10                  10.00
11                  12.00
12                  13.00
13                  16.00     11.67     12.17
14                  19.00     13.67     14.33
15                  23.00     16.00     17.00
16                  26.00     19.33     20.50
17                  30.00     22.67     23.83
18                  28.00     26.33     27.50
19                  18.00     28.00     28.33
20                  16.00     25.33     23.33
21                  14.00     20.67     18.67
```

Figure 1.17: *Three-Point Moving Average Spreadsheet*

```
/GF$
B3:THREE POI      C3:NT MOVING      D3:" AVERAGE
B5:ENTER DAT      C5:A
A7:/ - *
/R:B7.F7
B9:/FR DATA
C9:/FR UNWTD
   C13:(B10 + B11 + B12)/3
   /R:C14.C21:RRR
D9:/FR WTD
   D13:((3 * B12) + (2 * B11) + B10)/6
   /R:D14.D21:RRR
```

Figure 1.18: *Spreadsheet Instructions: Three-Point Moving Average Spreadsheet*

LINEAR REGRESSION

PURPOSE

This spreadsheet will perform a linear regression on a series of data points. The spreadsheet calculates the coefficients of the equation:

$$y = Mx + B$$

along with the correlation coefficient *R*, and a series of intermediate quantities.

	A	B	C	D	E	F
1						
2						
3		LINEAR REGRESSION				
4						
5		ENTER DATA				
6						
7	**					
8						
9		X	Y	X^2	Y^2	X*Y
10		1	100	1	10000	100
11		2	103	4	10609	206
12		3	108	9	11664	324
13		4	116	16	13456	464
14		5	120	25	14400	600
15		6	126	36	15876	756
16		7	129	49	16641	903
17		8	136	64	18496	1088
18		9	140	81	19600	1260
19		10	146	100	21316	1460
20		11	150	121	22500	1650
21		12	155	144	24025	1860
22						
23		N	12			
24		YAV	127.4167			
25		XAV	6.5			
26		E	732.5			
27		F	143.0000			
28		M	5.122378			
29		B	94.12121			
30		G	3762.917			
31		R	.9971365			

Figure 1.19: *Linear Regression Spreadsheet*

DATA ENTRY

You can enter up to twelve pairs of x and y data points. You may easily modify the spreadsheet to accommodate more points.

EXAMPLE

A sample regression analysis is displayed in Figure 1.19 (shown on the opposite page); the instructions for this spreadsheet are presented in Figure 1.20.

```
/GOR
B3:LINEAR RE        C3:GRESSION
B5:ENTER DAT        C5:A
A7:/ - *
/R:B7.F7
B9:/FR X
   B10:1
   B11:1 + B10
   /R:B12.B21:R
C9:/FR Y
D9:/FR X ^2
   D10: + B10 * B10
   /R:D11.D21:RR
E9:/FR Y ^2
   E10: + C10 * C10
   /R:E11.E21:RR
F9:/FR X * Y
   F10: + B10 * C10
   /R:F11.F21:RR
B23:N
   C23:@COUNT(B10.B21)
B24:YAV
   C24:@AVERAGE(C10.C21)
B25:XAV
   C25:@AVERAGE(B10.B21)
```

Figure 1.20: *Spreadsheet Instructions: Linear Regression*

```
B26:E
   C26:@SUM(F10.F21)-(@SUM(B10.B21)*@SUM(C10.C21)/C23)
B27:F
   C27:@SUM(D10.D21)-(@SUM(B10.B21)^2/C23)
B28:M
   C28:+C26/C27
B29:B
   C29:+C24-(C28*C25)
B30:G
   C30:@SUM(E10.E21)-(@SUM(C10.C21)^2/C23)
B31:R
C31:+C26*C26/C27/C30
```

Figure 1.20, Continued

CURVILINEAR REGRESSION

PURPOSE

This spreadsheet performs a curvilinear regression on a series of data points by using a transformation of variables and then using the linear technique that we developed in the previous example. This example has been set up with a series of seven x and y values; you may easily modify this example for your own data. The equation for which the data is fit is:

$$y = Bx^M$$

DATA ENTRY

You must enter the x and y data points. The spreadsheet will then compute the parameters of the curve and the correlation coefficient.

EXAMPLE

A sample spreadsheet appears in Figure 1.21. The spreadsheet instructions are shown in Figure 1.22.

```
        A         B           C          D          E          F          G          H
 1
 2
 3          CURVILINEAR REGRESSION
 4
 5          ENTER DATA
 6
 7  ****************************************************************************
 8
 9                    X           Y       LN X        LNY    LNX*LNY     LNX^2      LNY^2
10                 1000         300   6.907755   5.703782   39.40033   47.71708   32.53313
11                 1050         315   6.956545   5.752573   40.01803   48.39352   33.09209
12                 1100         330   7.003065   5.799093   40.61143   49.04293   33.62948
13                 1145         343   7.043160   5.837730   41.11607   49.60610   34.07910
14                 1190         356   7.081709   5.874931   41.60455   50.15060   34.51481
15                 1248         373   7.129298   5.921578   42.21669   50.82688   35.06509
16                 1310         392   7.177782   5.971262   42.86042   51.52056   35.65597
17
18          N                     7
19          YAV            5.837278
20          XAV            7.042759
21          E              .0536938
22          F              .0544720
23          M              .9857141
24          LN B           -1.10487   B          .3312544
25          G              .0529304
26          R              .9999300
```

Figure 1.21: *Curvilinear Regression Spreadsheet*

```
B3:CURVILINE      C3:AR REGRES      D3:SION
B5:ENTER DAT      C5:A
A7:/ – *
/R:B7.H7
B9:/FR X
C9:/FR Y
D9:/FR LN X
   D10:@LN(B10)
   /R:D11.D16:R
```

Figure 1.22: *Spreadsheet Instructions: Curvilinear Regression*

```
E9:/FR LNY
   E10:@LN(C10)
   /R:E11.E16:R
F9:/FR LNX*LNY
   F10:+D10*E10
   /R:F11.F16:RR
G9:/FR LNX^2
   G10:+D10*D10
   /R:G11.G16:RR
H9:/FR LNY^2
   H10:+E10*E10
   /R:H11.H16:RR
B18:N
   C18:@COUNT(B10.B16)
B19:YAV
   C19:@AVERAGE(E10.E16)
B20:XAV
   C20:@AVERAGE(D10.D16)
B21:E
   C21:@SUM(F10.F16)-(@SUM(D10.D16)*@SUM(E10.E16)/C18)
B22:F
   C22:@SUM(G10.G16)-(@SUM(D10.D16)^2/C18)
B23:M
   C23:+C21/C22
B24:LN B
   C24:+C19-(C23*C20)
D24:"   B
   E24:@EXP(C24)
B25:GB
   C25:@SUM(H10.H16)-(@SUM(E10.E16)^2/C18)
B26:R
   C26:+C21*C21/C22/C25
```

Figure 1.21, Continued

2

Electrical Engineering

DECIBEL CONVERSION

PURPOSE

This spreadsheet converts a set of voltage ratios into equivalent decibels and nepers. The following equations perform the conversions:

$$\text{dB} = 20 \log (V_2/V_1)$$

$$\text{neper} = .1151 \text{ dB}$$

DATA ENTRY

The spreadsheet provides the decibel and neper conversions for ten voltage ratios. You must enter the initial voltage ratio and the voltage ratio increment; the table is then generated from these parameters.

	A	B	C	D	E	F
1						
2						
3		DECIBEL CONVERSION				
4						
5		ENTER PARAMETERS:				
6						
7		VOLT RATIO		1.00		
8		RATIO INC		1.00		
9						
10	**					
11						
12		RATIO	DB	NEPER		
13		1.00	0.00	0.00		
14		2.00	6.02	0.69		
15		3.00	9.54	1.10		
16		4.00	12.04	1.39		
17		5.00	13.98	1.61		
18		6.00	15.56	1.79		
19		7.00	16.90	1.95		
20		8.00	18.06	2.08		
21		9.00	19.08	2.20		
22		10.00	20.00	2.30		

Figure 2.1: *Decibel Conversion Spreadsheet*

EXAMPLE

You wish to generate a table of decibel and neper conversions. Your table is to start with a voltage ratio of 1; the increment between voltage ratios is also 1. The conversion table is shown in Figure 2.1 (on the previous page); the spreadsheet instructions are presented in Figure 2.2 (below).

```
/GF$
B3:DECIBEL        C3:CONVERSIO    D3:N
B5:ENTER PAR      C5:AMETERS:
B7:VOLT RATI      C7:O
B8:RATIO INC
A10:/ - *
/R:B10.F10
B12:/FR RATIO
   B13: +D7
   B14: +D8+B13
   /R:B15.B22:NR
C12:/FR DB
   C13:20*@LOG10(B13)
   /R:C14.C22:R
D12:/FR NEPER
   D13:.1151*C13
   /R:D14.D22:R
```

Figure 2.2: *Spreadsheet Instructions: Decibel Conversion*

TEE ATTENUATOR

PURPOSE

An attenuator introduces a known loss between two resistive impedances to which the attenuator is matched. This spreadsheet develops a design table for a symmetrical Tee attenuator. The attenuator is inserted between equal impedances of value Z. The attenuator network as shown below consists of two series and one shunt resistor.

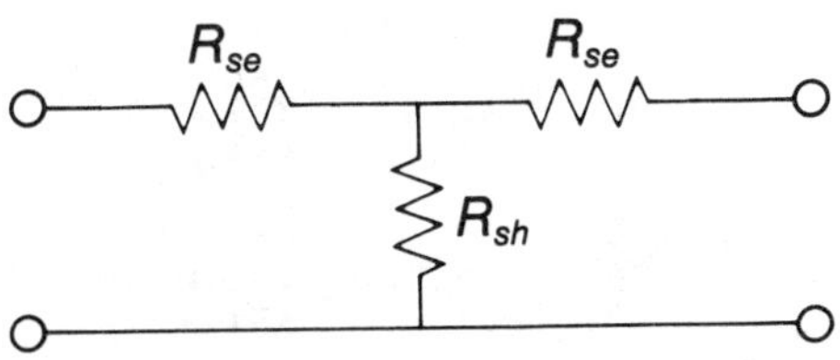

The design equations for the attenuator are:

$R_{sh} = Z/\sinh A$
$R_{se} = Z \tanh (A/Z)$
N = ratio of power absorbed
A = attenuation in Nepers = $.5 \ln N$

DATA ENTRY

You will enter three parameters into the spreadsheet. The first parameter, Z, are the impedances to which the attenuator is matched. You then enter an initial dB loss and a dB loss increment from which the design table is generated.

EXAMPLE

You wish to set up design tables for a Tee attenuator. The impedance which the attenuator matches is 500 ohms. The table should cover insertion losses in the range of one to ten dB. The design tables are shown in Figure 2.3; the instructions for the spreadsheet are displayed in Figure 2.4.

```
         A          B          C          D          E          F
 1
 2
 3            SYMMETRICAL T ATTENUATOR
 4
 5            ENTER PARAMETERS:
 6
 7            Z              500.00
 8            DB LOSS          1.00
 9            DB INC           1.00
10
11  ************************************************************
12
13               DBLOSS  SERIES R   SHUNT R
14                 1.00     28.74   4334.47
15                 2.00     57.30   2152.96
16                 3.00     85.48   1419.64
17                 4.00    113.11   1048.57
18                 5.00    140.03    822.64
19                 6.00    166.10    669.51
20                 7.00    191.19    558.20
21                 8.00    215.21    473.24
22                 9.00    238.06    406.05
23                10.00    259.69    351.49
```

Figure 2.3: *Tee Attenuator Spreadsheet*

```
/GF$
B3:SYMMETRIC     C3:AL T ATTE     D3:NUATOR
B5:ENTER PAR     C5:AMETERS:
B7:Z
B8:DB LOSS
B9:DB INC
A11:/-*
/R:B11.F11
B13:/FR DB LOSS
   B14:+C8
   B15:+B14+C9
   /R:B16.B23:RN
C13:/FR SERIES R
   C14:+C7*(1-@EXP(-2*.1151*B14/2))/(1+@EXP(-2*.1151*B14/2))
   /R:C15.C23:NRR
```

Figure 2.4: *Spreadsheet Instructions: Tee Attenuator*

```
D13:/FR SHUNT R
   D14: +C7/(@EXP(.1151*B14) - @EXP(-.1151*B14))*2
   /R:D15.D23:NRR
```

Figure 2.4, Continued

SYMMETRICAL PI ATTENUATOR

PURPOSE

The Pi attenuator is an alternative form of network to the Tee attenuator. This spreadsheet generates the design tables for a series of Pi attenuators based upon user-specified parameters. The design equations for the Pi attenuator shown below are:

$R_{sh} = Z / \tanh(A/2)$
$R_{se} = Z \sinh(A)$
N = ratio of power absorbed
A = attenuation in nepers $= .5 \ln N$

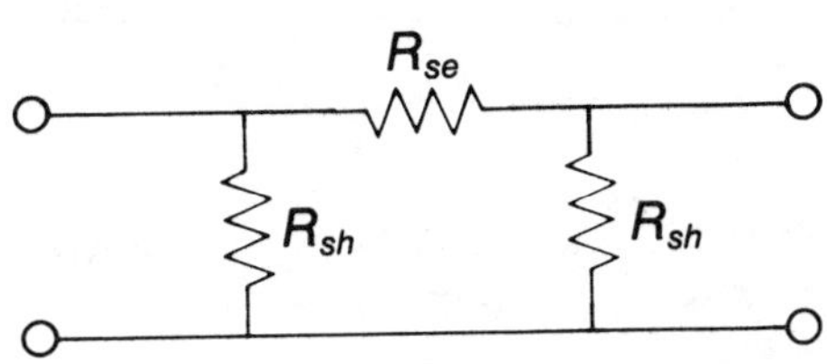

DATA ENTRY

You will enter three parameters into the spreadsheet. The first parameter, Z, is the impedance to which the attenuator is matched. You then enter an initial dB loss and dB loss increment from which the design table is generated.

EXAMPLE

You wish to set up design tables for a Pi attenuator. The impedance which the attenuator matches is 470 ohms. The table should cover insertion losses in the range of two to twenty dB. The design tables are shown in Figure 2.5; the instructions for the spreadsheet are displayed in Figure 2.6.

```
          A         B          C          D          E          F
 1
 2
 3            SYMMETRICAL PI ATTENUATOR
 4
 5            ENTER PARAMETERS:
 6
 7            Z                470.00
 8            DB LOSS            2.00
 9            DB INC             2.00
10
11  ***********************************************************************
12
13                     DB    SHUNT R   SERIES R
14                   2.00    4101.42     109.15
15                   4.00    2077.64     224.11
16                   6.00    1414.81     351.00
17                   8.00    1091.98     496.58
18                  10.00     904.91     668.58
19                  12.00     785.47     876.17
20                  14.00     704.42    1130.40
21                  16.00     647.14    1444.79
22                  18.00     605.46    1836.09
23                  20.00     574.51    2325.11
```

Figure 2.5: *Pi Attenuator Spreadsheet*

```
/GF$
B3:SYMMETRIC     C3:AL PI ATT     D3:ENUATOR
B5: ENTER PAR    C5:AMETERS:
B7:Z
B8:DB LOSS
B9:DB INC
A11:/- *
/R:B11.F11
B13:/FR DB
```

Figure 2.6: *Spreadsheet Instructions: Pi Attenuator*

```
B14: +C8
B15: +B14+C9
/R:B16.B23:RN
C13:/FR SHUNT R
C14: +C7/(1-@EXP(-2*.1151*B14/2))*(1+@EXP(-2*.1151*B14/2))
/R:C15.C23:NRR
D13:/FR SERIES R
D14: +C7*(@EXP(.1151*B14)-@EXP(-.1151*B14))/2
/R:D15.D23:NRR
```

Figure 2.6, Continued

AIR CORE INDUCTOR

PURPOSE

This spreadsheet displays a series of inductance values for an air core inductor. Given the geometric properties of the inductor, the spreadsheet calculates the inductance for different numbers of turns of wire.

The design equation for the air core inductor is:

$$L = n^2 r^2/(9r + 10l)$$

where:

L = inductance
n = number of turns
r = coil radius (inches), and
l = coil length (inches).

DATA ENTRY

You must specify four parameters prior to generating the spreadsheet. These parameters are the coil diameter, the spacing between turns, the initial number of turns, and the increment between turns in the design table. Note that coil length is equal to the number of turns multiplied by the turn spacing.

EXAMPLE

Let's produce design tables for a series of air core inductors. We will use a coil form 2 inches in diameter. Spacing between turns is .1 inches. Our table will begin with ten turns and increase in two-turn increments. The design table is displayed in Figure 2.7; the instructions are shown in Figure 2.8.

```
          A         B         C         D         E         F
 1
 2
 3             AIR CORE INDUCTOR
 4
 5             ENTER PARAMETERS:
 6
 7             COIL DIA       2.00
 8             TURN SPAC      0.10
 9             TURN          10.00
10             TURN INC       2.00
11
12  ************************************************************
13
14                  TURNS    INDUCT
15                  10.00      5.26
16                  12.00      6.86
17                  14.00      8.52
18                  16.00     10.24
19                  18.00     12.00
20                  20.00     13.79
21                  22.00     15.61
22                  24.00     17.45
23                  26.00     19.31
24                  28.00     21.19
```

Figure 2.7: *Air Core Inductor Spreadsheet*

```
/GF$
B3:AIR CORE        C3:INDUCTOR
B5:ENTER PAR       C5:AMETERS:
B7:COIL DIA
B8:TURN SPAC
B9:TURN
B10:TURN INC
A12:/ - *
/R:B12.F12
B14:/FR TURNS
   B15: +C9
   B16: +B15+C10
   /R:B17.B24:RN
C14:/FR INDUCT
   C15:(B15)^2*((C7/2)^2)/)(9*C7/2+(10*B15*C8))
   /R:C16.C24:RNNRN
```

Figure 2.8: *Spreadsheet Instructions: Air Core Inductor*

TRANSFORMER DESIGN

PURPOSE

You can use the VisiCalc program to create design tables for many types of transformers. This example shows you how to set up a spreadsheet to automate the transformer design. We use the following design equations in this example:

Secondary voltage:

$$E_S = \sqrt{(P_{\text{out}} Z_{\text{out}})}$$

Secondary current:

$$I_S = \frac{P_{\text{out}}}{E_S}$$

Turns ratio:

$$A = \sqrt{\frac{Z_{\text{in}}}{Z_{\text{out}}}}$$

Primary current:

$$I_P = 1.1 \frac{I_S}{A}$$

Primary turns:

$$N_P = \frac{E_P \times 10^8}{(4.44\, F A_c B_{\text{sat}})}$$

Primary inductance:

$$L = 1.25\,(N_p)^2\, \mu\, A_c\, E \times 10^{-8}/L_c$$

DATA ENTRY

You will enter a number of parameters for your particular design into the spreadsheet. These parameters include the input impedance, Z_{in}, the output impedance, Z_{out}, the output power, P_{out}, and the operating frequency, F. You also enter four core material parameters: the saturation flux density, the permeability, and the core area and length.

EXAMPLE

You wish to design a transformer which operates at 60 Hertz. The input and output impedances are 4000 and 10 ohms, respectively. The output power is 5 watts. You choose a core with an area of 6 centimeters, a length of 15 centimeters, a saturation flux density of 5000, and a permeability of 5000. The spreadsheet for this example is shown in Figure 2.9; the instructions for the setup of the spreadsheet are shown in Figure 2.10.

	A	B	C	D	E	F
1		TRANSFORMER DESIGN				
2						
3		ENTER PARAMETERS:				
4						
5		CORE AREA	6.00			
6		CORE LEN	15.00			
7		ZIN	4000.00			
8		ZOUT	10.00			
9		POUT	5.00			
10		FREQ	60.00			
11		BSAT	5000.00			
12		MU	5000.00			
13						
14	**					
15						
16		SEC VOLT		7.07		
17		SEC AMP		0.71		
18		TURNS RAT		20.00		
19		PRIM AMP		0.04		
20		PRIM VOLT		155.56		
21						
22						
23		PRIM TURN		1946.49		
24		SEC TURN		97.32		
25		INDUCT		94.72		

Figure 2.9: *Transformer Design Spreadsheet*

```
/GF$
B1:TRANSFORM   C1:ER DESIGN
B3:ENTER PAR   C3:AMETERS:
B5:CORE AREA
B6:CORE LEN
B7:ZIN
B8:ZOUT
B9:POUT
B10:FREQ
B11:BSAT
B12:MU
A14:/-*
/R:B14.F14
B16:SEC VOLT
  C16:(C8*C9)^.5
B17:SEC AMP
  C17:+C9/C16
B18:TURNS RAT
  C18:(C7/C8)^.5
B19:PRIM AMP
  C19:1.1*C17/C18
B20:PRIM VOLT
  C20:1.1*C18*C16
B23:PRIM TURN
  C23:1E8*C20/(4.44*C10*C5*C11)
B24:SEC TURN
  C24:+C23/C18
B25:INDUCT
  C25:1E-8*1.25*C23*C23*C12*C5/C6
```

Figure 2.10: *Spreadsheet Instructions: Transformer Design*

3 Electronic Engineering

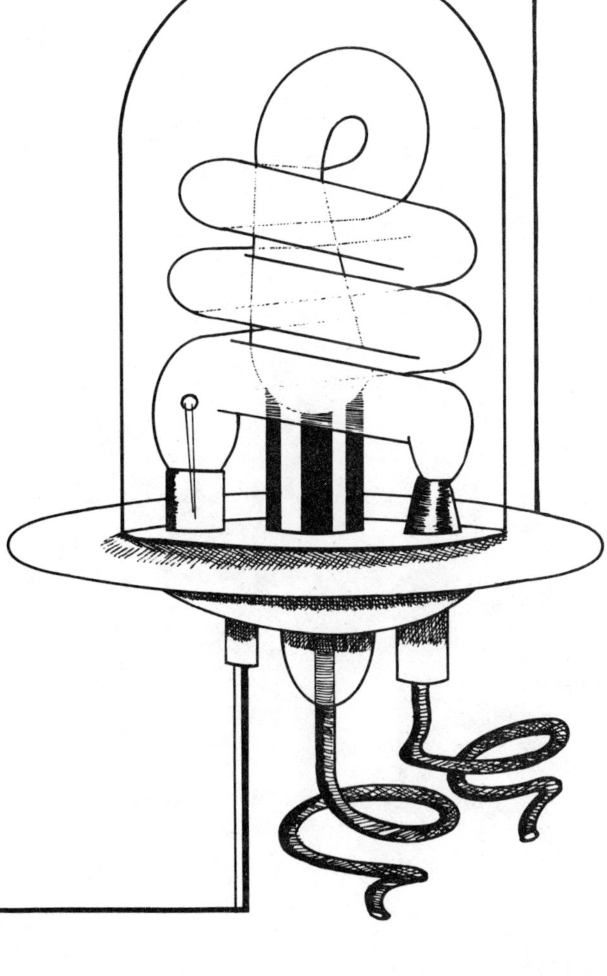

HIGH-PASS PASSIVE FILTER

PURPOSE

You can use this spreadsheet to find the circuit component values of both the Pi and Tee forms for a passive, constant-K, high-pass filter. The circuit diagrams and equations are shown below:

Pi form:

$$C = \frac{10^6}{4\pi fR}$$

$$L = \frac{10^3 R}{2\pi f}$$

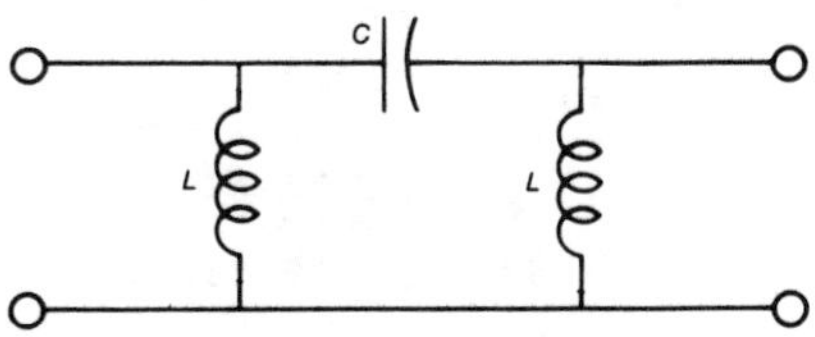

Tee form:

$$C = \frac{10^6}{2\pi fR}$$

$$L = \frac{10^3 R}{4\pi f}$$

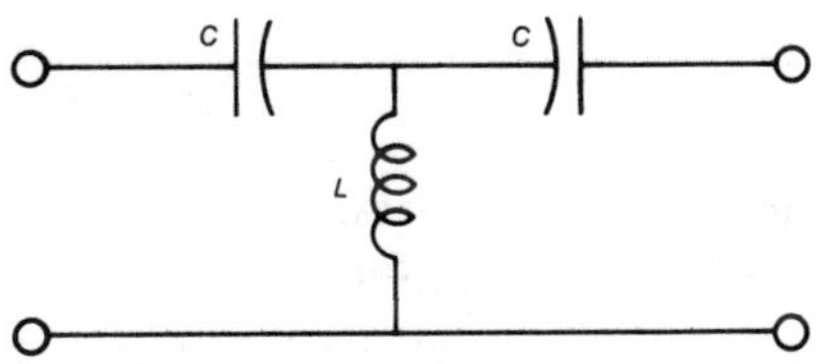

where

f = cutoff frequency, and
R= termination resistance.

DATA ENTRY

You enter the desired cutoff frequency and termination resistance. The spreadsheet calculates a design table which displays capacitance and inductance values for both forms of the filter.

EXAMPLE

You wish to find the design data for a high-pass filter having a cutoff frequency of 100 Hz and a termination resistance of 5Ω. Your design table is given in Figure 3.1, and your spreadsheet instructions are presented in Figure 3.2.

```
          A          B          C          D

1                   HIGH   PASS
2                PASSIVE   FILTER
3
4
5                ENTER PARAMETERS:
6
7                CUT. FREQ    100.00 HZ
8                TER. REST     50.00 OHMS
9
10   ************************************
11
12               CONSTANT K
13
14               HIGH PASS
15
16   VALUE             PI       TEE
17
18   CAP. C         15.92     31.83 MIC.F
19
20   IND. L         79.58     39.79 MIL.H
```

Figure 3.1: *High-Pass Passive Filter Spreadsheet*

```
/GF$
B1:"  HIGH          C1:PASS
B2:PASSIVE          C2:FILTER
B5:ENTER PAR        C5:AMETERS:
B7:CUT. FREQ        D7:" HZ
B8:TER. REST        D8:" OHMS
A10:/ - *
/R:B10.D10
B12:CONSTANT        C12:K
B14:HIGH PASS
A16:VALUE           B16:/FR PI          C16:/FR TEE
A18:CAP. C
   B18:1000000/(4*@PI*C7*C8)
   C18:1000000/(2*@PI*C8*C7)
   D18:" MIC.F
A20:IND. L
   B20:1000*C8/(2*@PI*C7)
   C20:1000*C8/(4*@PI*C7)
   D20:" MIL.H
```

Figure 3.2: *Spreadsheet Instructions: High-Pass Passive Filter*

LOW-PASS PASSIVE FILTER

PURPOSE

You can use this spreadsheet to find the circuit component values of both the Pi and Tee forms for a passive, constant-K, low-pass filter. The circuit diagrams and equations are:

Pi form:

$$C = \frac{10^6}{2\pi fR}$$

$$L = \frac{10^3 R}{\pi f}$$

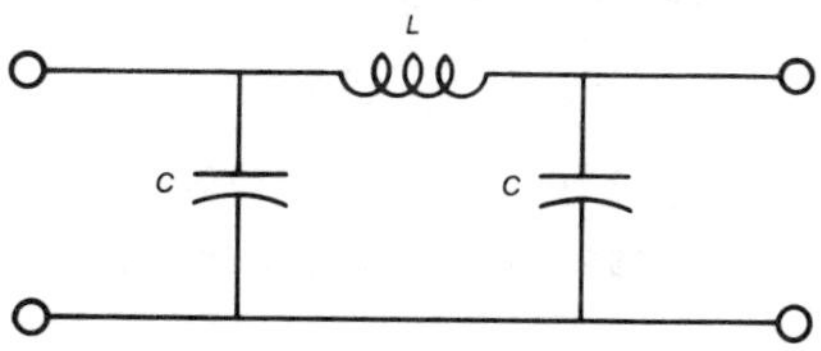

Tee form:

$$C = \frac{10^6}{\pi Rf}$$

$$L = \frac{10^3 R}{2\pi f}$$

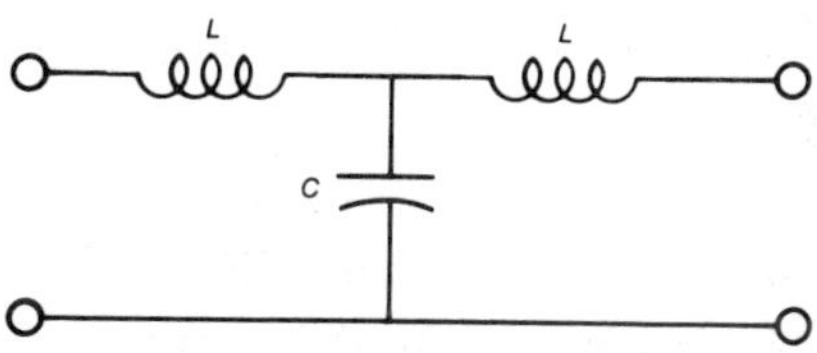

where

f = cutoff frequency, and
R= termination resistance.

DATA ENTRY

You enter two parameters: the desired cutoff frequency and termination resistance. The spreadsheet produces a design table which shows the capacitance and inductance values for both forms of the filter.

EXAMPLE

You wish to find the design data for a low-pass filter having a cutoff frequency of 100 Hz and a termination resistance of 50Ω. Your table is given in Figure 3.3, and your spreadsheet instructions are presented in Figure 3.4.

```
         A          B          C          D

1                   LOW   PASS
2              PASSIVE  FILTER
3
4
5              ENTER PARAMETERS:
6
7              CUT. FREQ     100.00 HZ
8              TER. REST      50.00 OHM
9
10 ****************************************
11
12             CONSTANT  K
13
14             LOW  PASS
15
16 VALUE              PI          TEE
17
18 CAP. C           31.83      63.66 MIC.F
19
20 IND. L          159.15      79.58 MIL.H
```

Figure 3.3: *Low-Pass Passive Filter Spreadsheet*

```
/GF$
B1:"  LOW          C1:PASS
B2:PASSIVE         C2:FILTER
B5:ENTER PAR       C5:AMETERS:
B7:CUT. FREQ       D7:"  HZ
B8:TER. REST       D8:"  OHM
A10:/ - *
/R:B10.D10
B12:CONSTANT       C12:K
B14:LOW PASS
A16:VALUE          B16:/FR PI          C16:/FR TEE
A18:CAP. C
   B18:1000000/(2*@PI*C7*C8)
   C18:1000000/(@PI*C8*C7)
   D18:"  MIC.F
A20:IND. L
   B20:1000*C8/(@PI*C7)
   C20:1000*C8/(2*@PI*C7)
   D20:"  MIL.H
```

Figure 3.4: *Spreadsheet Instructions: Low-Pass Passive Filter*

BAND-PASS PASSIVE FILTER

PURPOSE

This spreadsheet will compute the circuit component values of both the Pi and Tee forms for a passive, constant-K, band-pass filter. The circuit diagrams and equations are shown below:

Pi form:

$$C_1 = \frac{10^6(f_H - f_L)}{4\pi f_H f_L R}$$

$$C_2 = \frac{10^6}{2\pi(f_H - f_L)R}$$

$$L_1 = \frac{10^3 R}{\pi(f_H - f_L)}$$

$$L_2 = \frac{10^3 R(f_H - f_L)}{2\pi f_H f_L}$$

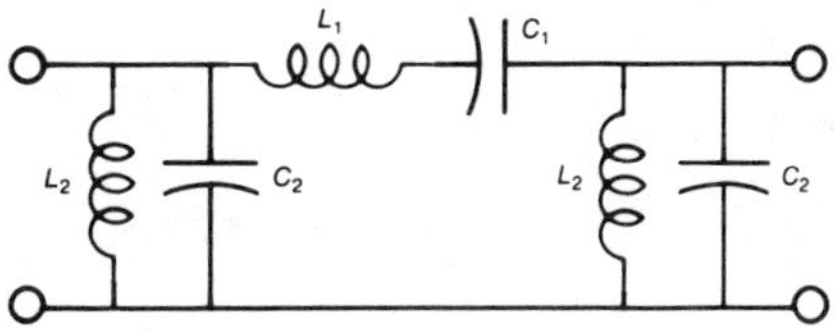

Tee form:

$$C_1 = \frac{10^6(f_H - f_L)}{2\pi f_H f_L R}$$

$$C_2 = \frac{10^6}{\pi(f_H - f_L)R}$$

$$L_1 = \frac{10^3 R}{2\pi(f_H - f_L)}$$

$$L_2 = \frac{10^3(f_H - f_L)R}{4\pi f_H f_L}$$

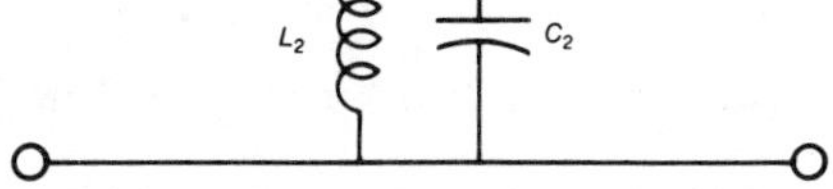

where:

f_L = low frequency cutoff,
f_H = high frequency cutoff, and
R = termination resistance.

DATA ENTRY

You enter the desired low and high cutoff frequencies and the desired termination resistance. The spreadsheet produces a design table which displays the capacitance and inductance values for both forms of the filter.

EXAMPLE

You wish to find the design data for a band-pass filter having a low cutoff frequency of 100 Hz, a high cutoff frequency of 500 Hz, and a termination resistance of 75Ω. Your design table is shown in Figure 3.5, and your spreadsheet instructions are presented in Figure 3.6.

```
          A          B          C          D
 1                  BAND    PASS
 2              PASSIVE    FILTER
 3
 4
 5              ENTER PARAMETERS:
 6
 7              LOW FREQ       100.00 HZ
 8              HI FREQ        500.00 HZ
 9              TER. REST       75.00 OHMS
10 ***************************************
11
12              CONSTANT K
13
14              BAND PASS
15
16 VALUE              PI          TEE
17
18 CAP. C1           8.49       16.98 MIC.F
19 CAP. C2           5.31       10.61 MIC.F
20 IND. L1          59.68       29.84 MIL.H
21 IND. L2          95.49       47.75 MIL.H
```

Figure 3.5: *Band-Pass Passive Filter Spreadsheet*

```
/GF$
B1:"   BAND        C1:PASS
B2:PASSIVE         C2:FILTER
B5:ENTER PAR       C5:AMETERS:
B7:LOW FREQ        D7:"  HZ
```

Figure 3.6: *Spreadsheet Instructions: Band-Pass Passive Filter*

```
B8:HI FREQ          D8:"  HZ
B9:TER. REST        D9:"  OHMS
A10:/ – *
/R:B10.D10
B12:CONSTANT        C12:K
B14:BAND PASS
A16:VALUE           B16:/FR PI          C16:/FR TEE
A18:CAP. C1
   B18:1000000*(C8 – C7)/(4*@PI*C7*C8*C9)
   C18:1000000*(C8 – C7)/(2*@PI*C8*C7*C9)
   D18:" MIC.F
A19:CAP. C2
   B19:1000000/(2*@PI*(C8 – C7)*C9)
   C19:1000000/(@PI*(C8 – C7)*C9)
   D19:" MIC.F
A20:IND. L1
   B20:1000*C9/(@PI*(C8 – C7))
   C20:1000*C9/(2*@PI*(C8 – C7))
   D20:" MIL.H
A21:IND. L2
   B21:1000*C9*(C8 – C7)/(2*@PI*C7*C8)
   C21:1000*C9*(C8 – C7)/(4*@PI*C7*C8)
   D21:" MIL.H
```

Figure 3.6, Continued

BAND-STOP PASSIVE FILTER

PURPOSE

You can use this spreadsheet to find the circuit component values of both the Pi and Tee forms for a passive, constant-K, band-stop filter. The circuit diagrams and equations are shown below:

Pi form:

$$C_1 = \frac{10^6}{4\pi(f_H - f_L)R}$$

$$C_2 = \frac{10^6(f_H - f_L)}{2\pi f_H f_L R}$$

$$L_1 = \frac{10^3(f_H - f_L)R}{\pi f_H f_L}$$

$$L_2 = \frac{10^3 R}{2\pi(f_H - f_L)}$$

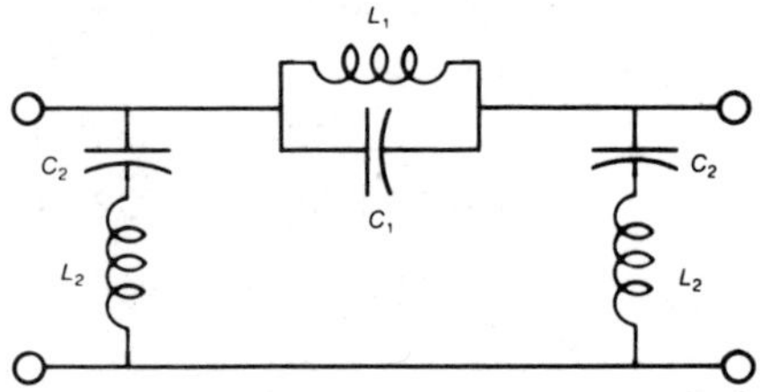

Tee form:

$$C_1 = \frac{10^6}{2\pi(f_H - f_L)R}$$

$$C_2 = \frac{10^6(f_H - f_L)}{\pi f_H f_L R}$$

$$L_1 = \frac{10^3(f_H - f_L)R}{2\pi f_H f_L}$$

$$L_2 = \frac{10^3 R}{4\pi(f_H - f_L)}$$

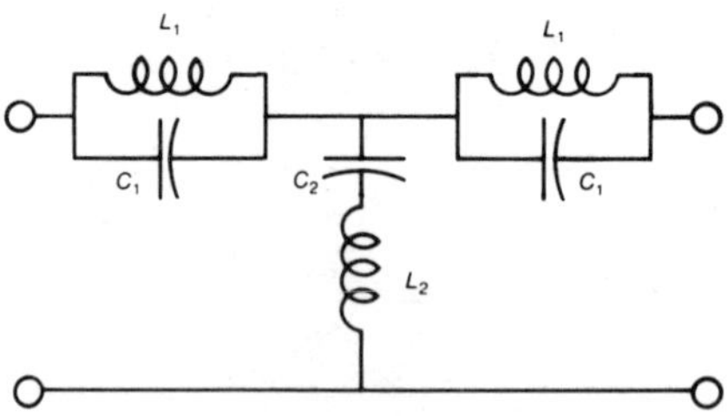

where:

f_L = low frequency cutoff,
f_H = high frequency cutoff, and
R = termination resistance.

DATA ENTRY

You enter the desired low and high cutoff frequencies and termination resistance. The spreadsheet produces a design table which shows the capacitance and inductance values for both forms of the filter.

EXAMPLE

You wish to find the design data for a band-stop filter having a low cutoff frequency of 100 Hz, a high cutoff frequency of 500 Hz, and a termination resistance of 75Ω. Your table is given in Figure 3.7, and your spreadsheet instructions are presented in Figure 3.8.

	A	B	C	D
1		BAND	STOP	
2		PASSIVE	FILTER	
3				
4				
5		ENTER PAR	AMETERS:	
6				
7		LOW FREQ	100.00	HZ
8		HI FREQ	500.00	HZ
9		TER. REST	75.00	OHMS
10	**********	**********	**********	********
11				
12		CONSTANT K		
13				
14		BAND STOP		
15				
16	VALUE	PI	TEE	
17				
18	CAP. C1	2.65	5.31	MIC.F
19	CAP. C2	16.98	33.95	MIC.F
20	IND. L1	190.99	95.49	MIL.H
21	IND. L2	29.84	14.92	MIL.H

Figure 3.7: *Band-Stop Passive Filter Spreadsheet*

```
/GF$
B1:"  BAND        C1:STOP
B2:PASSIVE        C2:FILTER
B5:ENTER PAR      C5:AMETERS:
B7:LOW FREQ       D7:"  HZ
```

Figure 3.8: *Spreadsheet Instructions: Band-Stop Passive Filter*

```
B8:HI FREQ          D8:" HZ
B9:TER. REST        D9:" OHMS
A10:/-*
/R:B10.D10
B12:CONSTANT        C12:K
B14:BAND STOP
A16:VALUE           B16:/FR PI          C16:/FR TEE
A18:CAP. C1
  B18:1000000/(4*@PI*(C8-C7)*C9)
  C18:1000000/(2*@PI*(C8-C7)*C9)
  D18:" MIC.F
A19:CAP. C2
  B19:1000000*(C8-C7)/(2*@PI*C8*C7*C9)
  C19:1000000*(C8-C7)/(@PI*C8*C7*C9)
  D19:" MIC.F
A20:IND. L1
  B20:1000*(C8-C7)*C9/(@PI*C8*C7)
  C20:1000*(C8-C7)*C9/(2*@PI*C8*C7)
  D20:" MIL.H
A21:IND. L2
B21:1000*C9/(2*@PI*(C8-C7))
C21:1000*C9/(4*@PI*(C8-C7))
D21:" MIL.H
```

Figure 3.8, Continued

FIRST-ORDER DIGITAL FILTER

PURPOSE

This spreadsheet computes the coefficients of a first-order digital filter and also tabulates the output response of the filter to a step input. The form of the filter is:

$$y_k = Ay_{k-1} + Bu_{k-1}$$

where

$$A = e^{-(T/\tau)} \qquad B = (1 - A)$$

and

T	= sampling time period,
τ	= filter time constant,
u_{k-1}	is filter input data, and
y_k, y_{k-1}	are filter output data.

DATA ENTRY

You first enter the sampling time and the filter time constant. Next enter the value of the filter input prior to the step; i.e., the amplitude at minus zero time. Then you enter the value of the step input to the filter; i.e., the amplitude at plus zero time. The resulting design table shows the values for A and B as well as the tabulated filter output corresponding to each point up to 100 points. Note that the 100 points occupy column C from C22 to C122, and that only the values down to line 40 are shown here.

EXAMPLE

You wish to simulate a first-order lag effect on digital data similar to an RC filter on analog data. Design parameters are as follows: the sampling time equals 10, the filter time constant equals 200, the amplitude at minus zero equals 0 and the amplitude at plus zero equals 10. The spreadsheet results are displayed in the table of Figure 3.9. Instructions for the spreadsheet are given in Figure 3.10.

```
         A          B          C          D
 1                DIGITAL   FILTER
 2                FIRST     ORDER
 3
 4                ENTER PARAMETERS:
 5
 6                SAMPLE T       10.00
 7                FILT TC       200.00
 8                AMP-ZERO        0.00
 9                AMP+ZERO       10.00
10   ************************************
11
12                FILTER PARAMETERS
13
14                     A=         0.95
15
16                     B=         0.05
17
18                STEP RESPONSE
19
20                POINT #    OUT DATA
21
22                   0.00         0.00
23                   1.00         0.49
24                   2.00         0.95
25                   3.00         1.39
26                   4.00         1.81
27                   5.00         2.21
28                   6.00         2.59
29                   7.00         2.95
30                   8.00         3.30
31                   9.00         3.62
32                  10.00         3.93
33                  11.00         4.23
34                  12.00         4.51
35                  13.00         4.78
36                  14.00         5.03
37                  15.00         5.28
38                  16.00         5.51
39                  17.00         5.73
40                  18.00         5.93
```

Figure 3.9: *First-Order Digital Filter Spreadsheet*

```
/GF$
B1:DIGITAL        C1:FILTER
B2:FIRST          C2:ORDER
B4:ENTER PAR      C4:AMETERS:
B6:SAMPLE T
```

Figure 3.10: *Spreadsheet Instructions: First-Order Digital Filter*

```
B7:FILT TC
B8:AMP-ZERO
B9:AMP + ZERO
A10:/ - *
/R:B10.D10
B12:FILTER PA       C12:RAMETERS
B14:/FR A=
   C14:@EXP( - C6/C7)
B16:/FR B=
   C16:1 - C14
B18:STEP RESP     C18:ONSE
B20:POINT #
   B22:0
   B23: +B22+1
   /R:B24.B122:R
C20:/FR OUT DATA
   C22: +C8
   C23:(C14*C22)+(C16*C9)
   /R:C24.C122:NRNN
```

Figure 3.10, Continued

ACTIVE LOW-PASS FILTER

PURPOSE

This spreadsheet allows us to design a two-pole, low-pass, active filter. We specify the requirements of our application and interact with the spreadsheet table to modify the design as we proceed. Modifications are entered in rows 26–34. The circuit diagram is shown below.

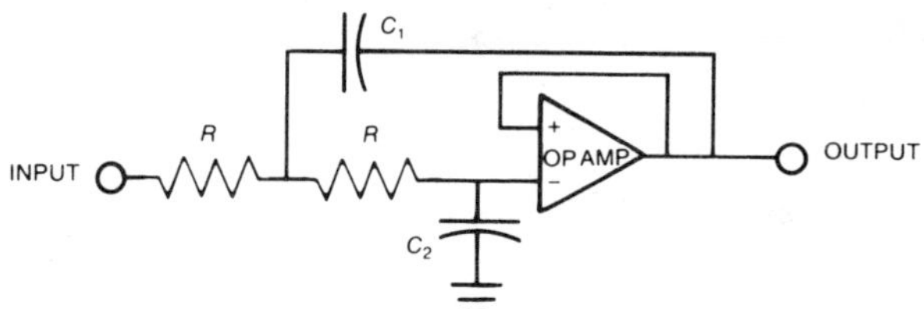

DATA ENTRY

We first enter the filter parameters: the circuit's Q factor, the ratio of the filter natural frequency to the −3 dB cutoff frequency, and the value of the −3 dB cutoff frequency. We then choose an initial design value for C_2. The design table produces the value for the filter natural frequency, and values for R and C_1. You may modify C_1, and the table will display new Q and R values. You may modify R, and a new value for the cutoff frequency will be generated.

EXAMPLE

We want to design a filter having a Q of 0.577, an F_n/F_c equal to 1.732, and a cutoff frequency of 10 Hz. We choose C_2 equal to 0.1 μf. The table shows the following: F_n equals 17.32 Hz, R equals 79.63 kΩ and C_1 equals 0.13317 μf. We decide to modify C_1 so C_1 equals 0.1332 μf. The spreadsheet table produces a new Q, equal to 0.57706 and a new R of 79.62 kΩ. We then choose a modified R, equal to 78.7 kΩ. Finally, a new cutoff frequency, F_c equals 10.12 Hz, is computed. These results are displayed on the spreadsheet table shown in Figure 3.11. Instructions for the spreadsheet are given in Figure 3.12.

```
         A           B           C          D
 1                   ACTIVE FILTER
 2                   (LOW-PASS)
 3
 4                   ENTER PARAMETERS:
 5
 6                   CKT Q.   =      .577
 7
 8                   FN/FC    =     1.732
 9
10                   CUT FREQ=         10 (HZ)
11
12                   CHOOSE C2         .1 (MIC. F)
13
14  ****************************************
15
16  FILTER PARAMETERS
17
18                   FN  =          17.32 (HZ)
19
20                   R   =       79.62812 (KOHMS)
21
22                   C1  =        0.13317 (MIC. F)
23
24  ****************************************
25
26                   ENTER MODIFICATIONS
27
28                   CHOOSE C1      .1332 (MIC.F)
29
30                   NEW Q =   .5770615
31
32                   NEW R =    79.6193 (KOHMS)
33
34                   CHOOSE R     78.70 (KOHMS)
35
36                   NEW FC =  10.11685 (HZ)
```

Figure 3.11: *Active Low-Pass Filter Spreadsheet*

```
B1:ACTIVE
C1:FILTER
B2:"  (LOW-          C2:PASS)
B4:ENTER PAR         D4:AMETERS:
B6:CKT Q. =
B8:FN/FC  =
B10:CUT. FREQ=       D10:" (HZ)
B12:CHOOSE C2        D12:" (MIC. F)
A14:/ - *
```

Figure 3.12: *Spreadsheet Instructions: Active Low-Pass Filter*

```
/R:B14.D14
A16:FILTER PA      B16:AMETERS
B18:FN =
   C18:+C8*C10
   D18:" (HZ)
B20:R  =
   C20:1000/(4*@PI*C12*C6*C18)
   D20:" (OHMS)
B22:C1 =
   C22:4*C6*C6*C12
   D22:" (MIC. F)
A24:/-*
/R:B24.D24
B26:ENTER MOD   C26:IFICATION       D26:S
B28:CHOOSE C1   D28:" (MIC.F)
B30:NEW Q =
   C30:.5*@SQRT(C28/C12)
B32: NEW R=
   C32:1000/(4*@PI*C12*C30*C18)
B34:CHOOSE R
   D34:" (KOHMS)
B36:NEW FC =
   C36:1000/(4*@PI*C30*C34*C8*C12)
   D36:" (HZ)
```

Figure 3.12, Continued

NON-INVERTING OPERATIONAL AMPLIFIER

PURPOSE

This spreadsheet calculates the circuit component values of a non-inverting operational amplifier. A multiplicative factor is used to provide alternate values for the feedback capacitor and hence, alternate resistor values which also meet the specified amplifier parameters. In addition, the amplifier zero frequency is calculated. The circuit and components are shown in the diagram below.

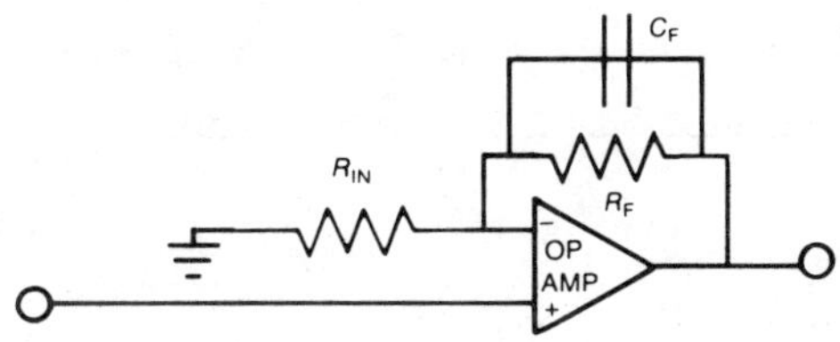

DATA ENTRY

You enter values for the amplifier gain, "pole" frequency, and feedback capacitor. A delta factor is also specified and used to multiply and divide the nominal capacitor value, as shown in the design table. Values for the feedback and input resistors are displayed for each of the three values of the feedback capacitor. The "zero" frequency value also appears in the table.

EXAMPLE

You are interested in an amplifier having a gain of 100 and a "pole" frequency of 100 Hz. You first try a capacitor of .01 μf. and choose a delta factor of 2. This gives alternate capacitor values of .005 and .02 μf. The table is displayed in Figure 3.13, and the spreadsheet instructions are shown in Figure 3.14.

```
            A           B           C           D
 1                  NON-INVERTING
 2                  OPERATIONAL AMP
 3
 4                  ENTER PARAMETERS:
 5
 6                  GAIN    =     100.00
 7                  FREQ(P)=      100.00 (HZ)
 8                  FB CAP =        0.01 (MIC. F)
 9                  DEL.   =        2.00
10
11 ****************************************
12
13                  FREQ (Z) = 10000.00 (HZ)
14
15                   CAP.*DEL        CAP   CAP/DEL
16
17 FB RES               79.58     159.15    318.31
18
19 INP RES               0.80       1.61      3.22
```

Figure 3.13: *Non-Inverting Operational Amplifier Spreadsheet*

```
/GF$
B1:NON-INVER      C1:TING
B2:OPERATION      C2:AL AMP.
B4:ENTER PAR      C4:AMETERS:
B6:GAIN     =
B7:FREQ(P)  =     D7:" (HZ)
B8:FB CAP   =     D8:" (MIC. F)
B9:DEL.     =
A11:/ - *
/R:B11.D11
B13:FREQ(Z) =
   C13: +C6*C7
   D13:" (HZ)
A17:FB RES      A19:INP RES
B15:CAP.*DEL
   B17:1000/(2*@PI*C8*C7*C9)
   B19: +B17/(C6 - 1)
C15:/FR CAP
```

Figure 3.14: *Spreadsheet Instructions: Non-Inverting Operational Amplifier*

```
    C17:1000/(2*@PI*C8*C7)
    C19: +C17/(C6-1)
D15:/FR CAP/DEL
    D17:1000*C9/(2*@PI*C8*C7)
    D19: +D17/(C6-1)
```

Figure 3.14, Continued

INVERTING OPERATIONAL AMPLIFIER

PURPOSE

You can use this spreadsheet to calculate the gain and cutoff frequency of an inverting operational amplifier stage. In addition, a sensitivity analysis is performed on a specified component. The circuit and components are shown in the diagram below.

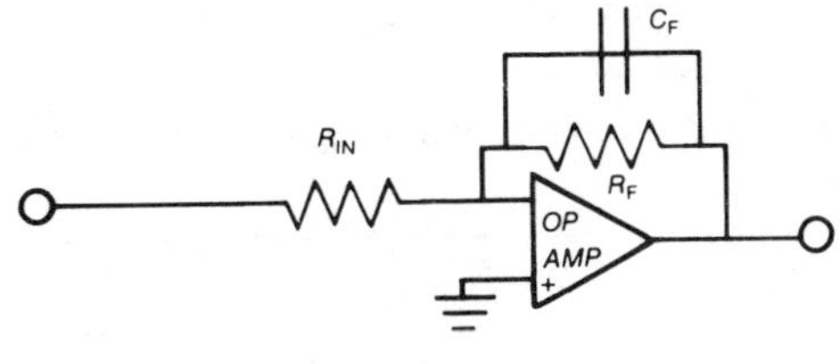

DATA ENTRY

You specify the input resistance and the feedback resistance and capacitance. Also, any one of those values may be chosen to be varied by a given percentage. This variance is entered as a delta value in percent. The design table then generates the gain and cutoff frequencies and a sensitivity analysis.

EXAMPLE

You wish to find the gain and cutoff frequency for an amplifier with an input resistor of 1 kΩ, a feedback resistor of 100 kΩ, and a feedback capacitor of .01 μf. Also, you wish to examine the effect of a $\pm 1\%$ and $\pm 2\%$ variation in the value of the feedback resistor. Hence, 1% is entered as the delta value for the feedback resistor. The table is illustrated in Figure 3.15, and the spreadsheet instructions are shown in Figure 3.16.

	A	B	C	D
1		INVERTING	OPERATIONAL	AMP
2				
3				
4		ENTER PARAMETERS:		
5				
6		R(IN)	1.00K	OHMS
7		DELR(IN)	0.00%	
8		R(FB)	100.00K	OHMS
9		DELR(FB)	1.00%	
10		C(FB)	0.01	MIC. F
11		DELC(FB)	0.00%	
12				
13	*********	*********	*********	*********
14				
15		#DEL	GAIN	F(P)
16		-2.00	98.00	162.40
17		-1.00	99.00	160.76
18		0.00	100.00	159.15
19		1.00	101.00	157.58
20		2.00	102.00	156.03

Figure 3.15: *Inverting Operational Amplifier Spreadsheet*

```
/GF$
B1:INVERTING       C1:" OPERATIO     D1:NAL AMP
B4:ENTER PAR       C4:AMETERS:
B6:R(IN) =         D6:K OHMS
B7:DELR(IN) =      D7:"%
```

Figure 3.16: *Spreadsheet Instructions: Inverting Operational Amplifier*

```
B8:R(FB) =          D8:K OHMS
B9:DELR(FB) =       D9:"%
B10:C(FB) =         D10:" MIC. F
B11:DELC(FB) =      D11:"%
A13:/ - *
/R:B13.D13
B15:"#DEL
   B16: -2
   B17:1 + B16
   /R:B18.B20:R
C15:/FR GAIN
   C16:(C8 * (1 + (B16 * C9/100))/(C6 * (1 + (B16 * C7/100)))
   /R:C17.C20:NRN NRN
D15:/FR F(P)
   D16:1000/(2 * @PI * C8 * (1 + (B16 * C9/100)) * C10 * (1 + (B16 * C11/100)))
   /R:D17.D20:NRN NRN
```

Figure 3.16, Continued

4

Communications

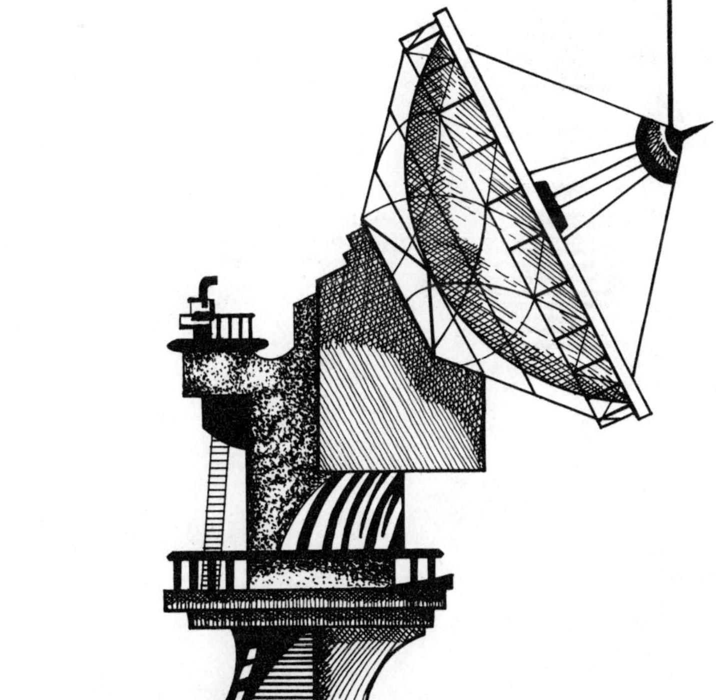

COAXIAL CABLE

PURPOSE

The spreadsheet in this example generates a table of characteristics of coaxial cables. The spreadsheet displays the capacitance, inductance, and characteristic impedance for arbitrary lengths of cable. The following equations are used in the computations:

capacitance:

$$\frac{2\pi\varepsilon}{\ln(R_1/R_2)}$$

inductance:

$$\mu/2\pi \times \ln R_1/R_2$$

impedance:

$$\sqrt{L/C}$$

The equations above yield the value per meter of cable, where ε and μ are the dielectric constant and permeability, respectively.

DATA ENTRY

Prior to generating the spreadsheet, you must specify the following quantities in metric units:

dielectric constant	E
permeability	U
inner radius	R1
outer radius	R2
starting length	LENGTH
length increment	LEN INC

EXAMPLE

Let's use the coaxial cable spreadsheet to calculate capacitance, inductance, and impedance for cables from 5 meters to 50 meters in length. The inner radius is .5 cm and the outer radius is 2 cm. The dielectric constant and permeability are 8.8E – 12 and 1.2E – 6, respectively. The spreadsheet and its instructions are shown in Figures 4.1 and 4.2.

```
         A          B          C          D          E          F
 1            COAXIAL CABLE
 2
 3            ENTER PARAMETERS:
 4
 5            U               .0000012
 6            E                8.8E-12
 7            R1                    .5
 8            R2                     2
 9            LENGTH                 5
10            LEN INC                5
11
12
13  *************************************************************
14
15            CAPAC/M    3.99E-11
16            INDUCT/M   2.648E-7
17            Z          81.47509
18
19               LENGTH     CAPAC    INDUCT
20                    5  1.99E-10  1.324E-6
21                   10  3.99E-10  2.648E-6
22                   15  5.98E-10  3.971E-6
23                   20  7.98E-10  5.295E-6
24                   25  9.97E-10  6.619E-6
25                   30  1.197E-9  7.943E-6
26                   35  1.396E-9  9.267E-6
27                   40  1.595E-9  1.059E-5
28                   45  1.795E-9  1.191E-5
29                   50  1.994E-9  1.324E-5
```

Figure 4.1: *Coaxial Cable Spreadsheet*

```
B1:COAXIAL        C1:CABLE
B3:ENTER PAR      C3:AMETERS:
B5:U              B6:E
B7:R1             B8:R2
B9:LENGTH         B10:LEN INC
A13:/ – *
/R:B13.F13
```

Figure 4.2: *Spreadsheet Instructions: Coaxial Cable*

```
B15:CAPAC/M
   C15:2*@PI*C6/@LN(C8/C7)
B16:INDUCT/M
   C16: +C5/2/@PI*@LN(C8/C7)
B17:Z
   C17:(C16/C15)^.5
B19:/FR LENGTH
   B20: +C9
   B21: +C10+B20
   /R:B22.B29:NR
C19:/FR CAPAC
   C20: +B20*C15
   /R:C21.C29:RN
D19:/FR INDUCT
   D20: +B20*C16
   /R:D21.D29:RN
```

Figure 4.2, Continued

WAVEGUIDE CALCULATIONS

PURPOSE

The spreadsheet we will develop in this example calculates the cutoff frequencies for several modes in a rectangular waveguide. The cutoff frequencies apply to both transverse electric (TE) and transverse magnetic (TM) modes in the guide. The equation for the cutoff frequency is given as:

$$F_c = \frac{1}{2} \times \frac{1}{\sqrt{\mu \varepsilon}} \times \sqrt{(M/A)^2 + (N/B)^2}$$

DATA ENTRY

You must specify the width of the guide, *A,* and the height of the guide, *B,* as well as the dielectric constant and permeability of the media in the guide. Additionally, you must enter the mode numbers, *M* and *N.*

EXAMPLE

Let's use the spreadsheet to calculate the cutoff frequencies for a 4 × 7 cm guide with TE and TM modes 0 through 2. The dielectric constant is 8.8E–12 and the permeability is 1.256E–6. The resulting spreadsheet is shown in Figure 4.3; the spreadsheet instructions appear in Figure 4.4.

```
         A          B          C          D          E          F
 1
 2
 3               WAVEGUIDE CALCULATIONS
 4
 5               ENTER PARAMETERS:
 6
 7               A                .04
 8               B                .07
 9               E            8.8E-12
10               U           1.256E-6
11
12
13
14   ************************************************************
15
16                          M         N        FC
17                          0         1  2.1485E9
18                          1         0  9.400E10
19                          1         1  9.615E10
20                          0         2  4.2970E9
21                          2         0  3.760E11
22                          1         2  9.829E10
23                          2         1  3.781E11
```

Figure 4.3: *Waveguide Spreadsheet*

```
B3:WAVEGUIDE     C3:" CALCULAT     D3:IONS
B5:ENTER PAR     C5:AMETERS:
B7:A             B8:B
```

Figure 4.4: *Spreadsheet Instructions: Waveguide Calculations*

```
B9:E                B10:U
A14:/ - *
/R:B14.F14
B16:/FR M           C16:/FR N
D16:/FR FC
   D17:.5/((C9*C10)^.5)*(((B17/C7)^2+(C17/C8)^2)^.5)
   /R:D18.D23:NNRNRN
```

Figure 4.4, Continued

MICROSTRIP TRANSMISSION LINE

PURPOSE

Microstrip transmission lines are used on printed-circuit boards to carry very fast signals. We will use the VisiCalc program to generate a table of parameters for microstrip lines. The impedance of the line is computed from the equation:

$$\frac{377H}{W\sqrt{E}}$$

DATA ENTRY

You must specify the following parameters prior to executing the spreadsheet: H, the distance between the conductor and the ground plane; ε, the relative dielectric constant of the medium; and the initial width and width increment for the line.

EXAMPLE

You are designing a printed-circuit board and require the use of a micro-strip line. The board thickness is 5 mils and the relative dielectric constant is 3.1. You are considering line widths from 5 to 35 mils. Figure 4.5 shows the resulting design spreadsheet. The spreadsheet instructions appear in Figure 4.6.

```
        A         B         C         D         E         F
 1
 2
 3                MICROSTRIP TRANSMISSION LINE
 4
 5                ENTER PARAMETERS:
 6
 7                H                  5.00
 8                E                  3.10
 9                WIDTH              5.00
10                WID INC            5.00
11
12
13
14  *****************************************************
15
16                     WIDTH     IMPED
17                      5.00    214.12
18                     10.00    107.06
19                     15.00     71.37
20                     20.00     53.53
21                     25.00     42.82
22                     30.00     35.69
23                     35.00     30.59
```

Figure 4.5: *Microstrip Transmission Line Spreadsheet*

```
/GF$
B3:MICROSTRI      C3:P TRANSMI
D3:SSION LIN      E3:E
B5:ENTER PAR      C5:AMETERS:
B7:H              B8:E
B9:WIDTH
B10:WID INC
A14:/-*
```

Figure 4.6: *Spreadsheet Instructions: Microstrip Transmission Line*

```
/R:B14.F14
B16:/FR WIDTH
   B17: +C9
   B18: +C10+B17
   /R:B19.B23:NR
C16:/FR IMPED
   C17:377*C7/B17/(C8^.5)
   /R:C18.C23:NRN
```

Figure 4.6, Continued

PARABOLIC ANTENNA

PURPOSE

Let's use the VisiCalc program to calculate the gain and bandwidth of a parabolic antenna at various operating frequencies. The equations for the gain and bandwidth are:

$$\text{Gain} = 10 \log (10.2E) + 20 \log D + 20 \log f$$

$$B_w = \frac{66.4}{f/D}$$

where:

D = diameter,
f = frequency, and
E = efficiency.

DATA ENTRY

We set up a spreadsheet which requires the entry of the antenna efficiency, the antenna diameter, the operating frequency, and the frequency increment to be used in the table.

EXAMPLE

You wish to make antenna calculations for an antenna operating at frequencies ranging from 7.4 GHz to 9.2 GHz. The antenna diameter is 12 ft and efficiency is 55%. The resulting spreadsheet is displayed in Figure 4.7, and its specifications in Figure 4.8.

```
           A         B          C          D          E          F
 1
 2
 3               PARABOLIC  ANTENNA
 4
 5               ENTER  PARAMETERS:
 6
 7               DIAM.           12.00
 8               FREQ.            7.40
 9               FREQ  INC.       0.20
10               EFFIC.           0.55
11
12  *****************************************************************
13
14
15                     FREQ       GAIN   BW(3DB)
16                     7.40      46.46      0.75
17                     7.60      46.69      0.73
18                     7.80      46.92      0.71
19                     8.00      47.14      0.69
20                     8.20      47.35      0.67
21                     8.40      47.56      0.66
22                     8.60      47.76      0.64
23                     8.80      47.96      0.63
24                     9.00      48.16      0.61
25                     9.20      48.35      0.60
```

Figure 4.7: *Parabolic Antenna Spreadsheet*

```
/GF$
B3:PARABOLIC      C3:" ANTENNA
B5:ENTER PAR      C5:AMETERS:
B7:DIAM.
B8:FREQ.
B9:FREQ INC.
B10:EFFIC.
A12:/- *
```

Figure 4.8: *Spreadsheet Instructions: Parabolic Antenna*

```
/R:B12.F12
B15:/FR FREQ
   B16: +C8
   B17: +B16+C9
   /R:B18.B25:RN
C15:/FR GAIN
   /C16:@LOG10(C7)*20+(10*@LOG10(10.2*C10))+(20*@LOG10(B16))
   /R:C17.C25:NNR
D15:/FR BW(3DB)
   D16:66.4/B16/C7
   /R:D17.D25:RN
```

Figure 4.8, Continued

LOADED DIPOLE ANTENNA

PURPOSE

We can set up a design spreadsheet to calculate the loading inductance for resonance for dipole antennas. Given the overall length of the antenna, the length from the center to the coil, the operating frequency, and the wire diameter, we can calculate the required inductance. A diagram of the antenna appears below.

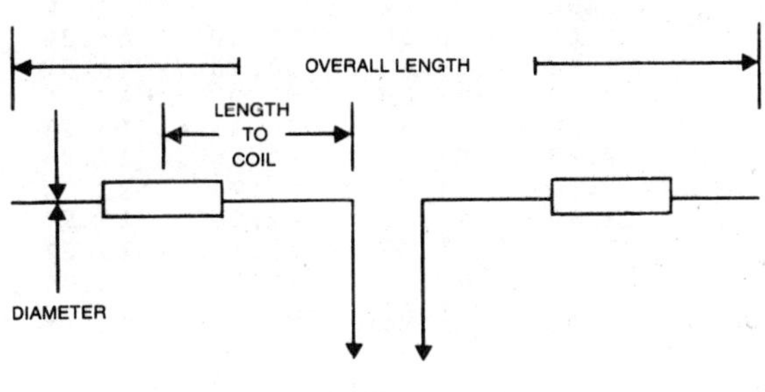

DATA ENTRY

You will enter the antenna length in feet, the length to the coil in feet, and the wire diameter in inches. In order to generate a series of inductances, you also enter an initial frequency and a frequency increment.

EXAMPLE

Let's design a coil to operate at 1.8 MHz. We can then examine a series of operating frequencies increasing in steps of 0.1 MHz. The length of the dipole antenna is 120 feet and the coil will be located 15 feet from the center. The diameter of the wire is 0.12 in. The spreadsheet for this example is shown in Figure 4.9 and its instructions in Figure 4.10. Note that a number of intermediate calculations are also displayed.

```
        A         B         C         D         E         F
 1
 2
 3            LOADING INDUCTANCE FOR
 4            DIPOLE ANTENNA
 5
 6            ENTER PARAMETERS:
 7
 8            OVERALL LENGTH          120.00
 9            LENGTH TO COIL           15.00
10            FREQUENCY                 1.80
11            FREQUENCY INC             0.10
12            DIAMETER                  0.12
13
14  ************************************************************
15
16            X=             45.00
17            W=            200.00
18
19                  FREQ         Z         Y         L
20                  1.80    130.00    115.00     65.04
21                  1.90    123.16    108.16     56.57
22                  2.00    117.00    102.00     49.35
23                  2.10    111.43     96.43     43.13
24                  2.20    106.36     91.36     37.74
25                  2.30    101.74     86.74     33.04
26                  2.40     97.50     82.50     28.91
27                  2.50     93.60     78.60     25.27
28                  2.60     90.00     75.00     22.04
29                  2.70     86.67     71.67     19.16
```

Figure 4.9: *Loaded Dipole Antenna Spreadsheet*

```
/GF$
B3:LOADING I        C3:NDUCTANCE
D3:" FOR            B4:DIPOLE AN
C4:TENNA
B6:ENTER PAR        C6:AMETERS:
B8:OVERALL L        C8:ENGTH
B9:LENGTH TO        C9:" COIL
B10:FREQUENCY
B11:FREQUENCY  C11:" INC
B12:DIAMETER
A14:/ - *
/R:B14.F14
B16:X =
   C16: + D8/2 - D9
B17:W =
   C17:24/D12
B19:/FR
/R:C19.E19
B19:FREQ
   B20: + D10
   B21: + B20 + D11
   /R:B22.B29:RN
C19:Z
   C20:234/B20
   /R:C21.C29:R
D19:Y
   D20: + C20 - D9
   /R:D21.D29:RN
E19:L
   E20:1000000/(68 * ((@PI * B20) ^2)) * ((@LN(C17 * D20) - 1) *
      ((D20/C20) ^2 - 1)/D20 - (((C16/C20) ^2 - 1) *
      (@LN(C16 * C17) - 1)/C16))
   /R:E21.E29:RNR RRR NRN NN
```

Figure 4.10: *Spreadsheet Instructions: Loaded Dipole Antenna*

LOADED VERTICAL ANTENNA

PURPOSE

This application generates a spreadsheet for calculating the loading inductance for a vertical antenna. A diagram of the antenna is shown below.

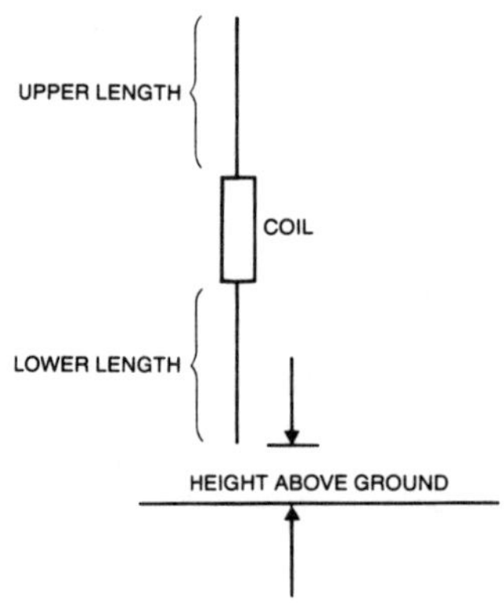

DATA ENTRY

We have set up the spreadsheet so that you specify the lengths of the lower and upper sections, the height of the antenna above ground, the average thickness of the antenna, all in inches, and the desired operating frequency.

EXAMPLE

You wish to examine the required loading inductances for frequencies from 6.8 to 7.6 MHz. The upper and lower lengths of the antenna are 60 and 50 inches. The antenna is 20 inches above ground. Its average thickness is .15 inches. The spreadsheet for this example appears in Figure 4.11 and its instructions in Figure 4.12. Note that the spreadsheet displays a series of intermediate quantities, the inductance L, and the radiation resistance RRAD.

```
         A         B         C         D         E         F
 1
 2
 3                 LOADED VERTICAL ANTENNA
 4
 5                 ENTER PARAMETERS:
 6
 7                 HT ABOVE GROUND          20.00
 8                 LOWER     LENGTH         60.00
 9                 UPPER     LENGTH         50.00
10                 AVG THICKNESS             0.15
11                 FREQ                      6.80
12                 FREQ INC                  0.10
13
14 ****************************************************
15
16                 ZO        405.42
17
18                      FREQ      BDEG      CDEG         L      RRAD
19                      6.80     12.44     10.36     49.80      3.66
20                      6.90     12.62     10.52     48.29      3.76
21                      7.00     12.80     10.67     46.84      3.87
22                      7.10     12.98     10.82     45.46      3.98
23                      7.20     13.17     10.97     44.13      4.08
24                      7.30     13.35     11.12     42.85      4.20
25                      7.40     13.53     11.28     41.63      4.31
26                      7.50     13.72     11.43     40.46      4.42
27                      7.60     13.90     11.58     39.33      4.53
```

Figure 4.11: *Loaded Vertical Antenna Spreadsheet*

```
/GF$
B3:LOADED VE      C3:RTICAL AN      D3:TENNA
B5:ENTER PAR      C5:AMETERS:
B7:HT ABOVE       C7:GROUND
B8:LOWER          C8:LENGTH
B9:UPPER          C9:LENGTH
B10:AVG THICK     C10:NESS
B11:FREQ
B12:FREQ INC
A14:/ – *
/R:B14.F14
B16:ZO
   C16:138 * @LOG10(@SUM(D7.D9)/D10)
```

Figure 4.12: *Spreadsheet Instructions: Loaded Vertical Antenna*

```
B18:/FR
/R:C18.F18
B18:FREQ
   B19: +D11
   B20: +D12+B19
   /R:B21.B27:NR
C18:BDEG
   C19: +D8*B19/32.81
   /R:C20.C27:NR
D18:CDEG
   D19: +D9*B19/32.81
   /R:D20.D27:NR
E18:L
   E19:1/@TAN(@PI/180*D19) - @TAN(@PI/180*C19)*C16/(2*@PI*B19)
   /R:E20.E27:RRNR
F18:RRAD
   F19:(D19*@COS(@PI/180*C19)/2+
      (C19*(1+@COS(@PI/180*C19))/2)^2/82.3
   /R:F20.F27:RRRR
```

Figure 4.12, Continued

5 Solar Engineering

SOLAR ECONOMIC ANALYSIS

PURPOSE

This example will show you how to use the VisiCalc program to study the economics of installing solar equipment. We will account for immediate fuel savings and project the rising cost of fuel in doing our study. The VisiCalc program provides a net present value function; we use this function to perform the analysis.

DATA ENTRY

You will enter several variables into the spreadsheet prior to doing the economic analysis: the projected initial fuel savings, the percentage by which fuel prices are expected to escalate, the initial cost of the equipment, the annual maintenance rate, and the interest rate at which your funds are invested.

EXAMPLE

You wish to determine the economic feasibility of installing solar heating equipment. The equipment will cost $4500 and will save $750 per year in heating costs. Annual maintenance is expected to cost $50 per year. You expect fuel costs to rise 12% annually; unused funds may be invested at 11% per annum. The spreadsheet for this example is shown in Figure 5.1, and the spreadsheet instructions in Figure 5.2.

```
          A          B          C          D          E          F
 1
 2
 3             SOLAR ECONOMIC ANALYSIS
 4
 5             ENTER PARAMETERS:
 6
 7             INIT FUEL SAVINGS       750.00
 8             FUEL ESC RATE            12.00
 9             EQUIP COST             4500.00
10             MAINT                    50.00
11            INTEREST RATE            11.00
12
13 ****************************************************************
14
15                    YEAR     SAVED       MAINT       NET
16                    1.00    750.00       50.00    700.00
17                    2.00    840.00       50.00    790.00
18                    3.00    940.80       50.00    890.80
19                    4.00   1053.70       50.00   1003.70
20                    5.00   1180.14       50.00   1130.14
21                    6.00   1321.76       50.00   1271.76
22                    7.00   1480.37       50.00   1430.37
23                    8.00   1658.01       50.00   1608.01
24                    9.00   1856.97       50.00   1806.97
25                   10.00   2079.81       50.00   2029.81
26
27
28                     NPV   2242.90
```

Figure 5.1: *Solar Economic Analysis Spreadsheet*

```
/GF$
/GOR
B3:SOLAR ECO      C3:NOMIC ANA      D3:LYSIS
B5:ENTER PAR      C5:AMETERS:
B7:INIT FUEL      C7:" SAVINGS
B8:FUEL ESC       C8:RATE
B9:EQUIP COS      C9:T
B10:MAINT
B11:INTEREST      C11:RATE
A13:/ - *
/R:B13.F13
B15:/FR
```

Figure 5.2: *Spreadsheet Instructions: Solar Economic Analysis*

```
/R:C15.E15
B15:YEAR
   B16:1
   B17:1+B16
   /R:B18.B25:R
C15:SAVED
   C16:+D7
   C17:+C16*(1+(D8/100))
   /R:C18.C25:RN
D15:MAINT
   D16:+D10
   /R:D17.D25:N
E15:NET
   E16:+C16-D16
   /R:E17.E25:RR
B28:/FR NPV
   C28:-D9+@NPV(D11/100,E16.E25)
```

Figure 5.2, Continued

ANGLE OF DECLINATION

PURPOSE

This spreadsheet calculates the angle of declination of the sun for various days of the year. The first day of the solar year is December 21. For any day of the solar year, the angle of declination is given by the following equation:

$$\sin d = -\cos \frac{180(D_s - 1)}{182.6} \sin (23.45)$$

DATA ENTRY

You must enter two parameters into the solar declination spreadsheet: the number of the first day of interest and the increment in days. The spreadsheet will then generate the angle of declination for a series of ten points at the specified interval.

EXAMPLE

You wish to have the VisiCalc program list the angle of declination of the sun starting with December 21 and continuing in five-day increments. The spreadsheet is shown in Figure 5.3. The instructions for this spreadsheet are in Figure 5.4.

```
        A         B         C         D         E         F
 1
 2
 3            ANGLE OF DECLINATION
 4
 5            FIRST DAY OF SOLAR YEAR IS DEC. 21
 6
 7            ENTER PARAMETERS:
 8
 9            DAY NO.             1
10            DAY INC.            5
11
12  ************************************************************
13
14                  DAY      ANGLE
15
16                    1     -23.45
17                    6     -23.36
18                   11     -23.09
19                   16     -22.63
20                   21     -22.00
21                   26     -21.21
22                   31     -20.25
23                   36     -19.15
24                   41     -17.90
25                   46     -16.53
```

Figure 5.3: *Angle of Declination Spreadsheet*

```
B3:ANGLE OF       C3:DECLINATI      D3:ON
B5:FIRST DAY      C5:" OF SOLAR     D5:" YEAR IS      E5:DEC. 21
B7:ENTER PAR      C7:AMETERS:
B9:DAY NO.
B10:DAY INC.
A12:/ - *
/R:B12.F12
B14:/FR DAY
   B16: + C9
   B17: + B16 + C10
   /R:B18.B25:RN
C14:/FR ANGLE
   C16:/F$
   C16:180/@PI * @ASIN( - .398 * @COS((B16 - 1) * .986 * @PI/180))
   /R:C17.C25:R
```

Figure 5.4: *Spreadsheet Instructions: Angle of Declination*

SOLAR ALTITUDE AND AZIMUTH

PURPOSE

The spreadsheet in this example gives the solar altitude and azimuth at various hours of the day for a particular latitude and angle of declination. The equations used by the spreadsheet are:

Solar Altitude:

$$\sin(\text{alt}) = \sin L \sin d + \cos L \cos d \cos H_s$$

Solar Azimuth:

$$\sin(\text{az}) = \cos d \sin H_s / \cos(\text{az})$$

where:

L = latitude,
d = declination, and
H_s = solar hour.

DATA ENTRY

You will enter the latitude and the angle of declination for the day of interest. The spreadsheet will then calculate the solar altitude and azimuth at different hours during the day.

EXAMPLE

You need to calculate the solar altitude and azimuth for a site at a latitude of 40 degrees. The angle of declination for the day you are reviewing is −23 degrees. The resulting spreadsheet and its instructions are presented in Figures 5.5 and 5.6.

```
         A         B         C         D         E         F
 1
 2
 3            SOLAR ALTITUDE AND AZIMUTH
 4
 5            ENTER PARAMETERS:
 6
 7            LATITUDE       40.00
 8            DECLIN        -23.00
 9
10  ************************************************************
11
12            SIN(D)         -0.39
13            COS(D)          0.92
14            SIN(L)          0.64
15            COS(L)          0.77
16
17                   HOUR      ALT        AZ
18                 800.00     5.82    -53.26
19                 900.00    14.33    -42.21
20                1000.00    21.07    -29.55
21                1100.00    25.47    -15.30
22                1200.00    27.00      0.00
23                1300.00    25.47     15.30
24                1400.00    21.07     29.55
25                1500.00    14.33     42.21
26                1600.00     5.82     53.26
```

Figure 5.5: *Solar Altitude and Azimuth Spreadsheet*

```
/GF$
B3:SOLAR ALT      C3:ITUDE AND      D3: AZIMUTH
B5:ENTER PAR      C5:AMETERS:
B7:LATITUDE
B8:DECLIN
A10:/ – *
/R:B10.F10
B12:SIN(D)
   C12:@SIN(@PI/180*C8)
B13:COS(D)
   C13:@COS(@PI/180*C8)
B14:SIN(L)
   C14:@SIN(@PI/180*C7)
```

Figure 5.6: *Spreadsheet Instructions: Solar Altitude and Azimuth*

```
B15:COS(L)
   C15:@COS(@PI/180*C7)
B17:/FR
/R:C17.D17
B17:HOUR
   B18:800
   B19:100+B18
   /R:B20.B26:R
C17:ALT
   C18:180/@PI*@ASIN(C12*C14+(C13*C15
       *@COS(@PI/180*(B18/100-12)*15))
   /R:C19.C26:NNNNR
D17:AZ
   D18:180/@PI*@ASIN(C13/@COS(@PI/180*C18)*
       @SIN(@PI/180*(B18/100-12)*15))
   /R:D19.D26:NRR
```

Figure 5.6, Continued

SOLAR RAYS ON TILTED SURFACE

PURPOSE

Solar collectors are usually installed at an angle to the horizontal plane. This spreadsheet will calculate the angle of incidence for a tilted surface. The incidence angle is calculated from the equation:

$$\cos i = \sin(L - T) \sin d + \cos (L - T) \cos d \cos H_s$$

DATA ENTRY

You must first set up the spreadsheet and then enter the latitude, tilt of the collector, and angle of declination of the sun. The spreadsheet then calculates several intermediate quantities and the incidence angle for various hours of the day.

EXAMPLE

A collector is mounted with a tilt of 35 degrees at a latitude of 35 degrees. You wish to calculate the incident radiation on a day for which the angle of declination is 18 degrees. The spreadsheet for this example is displayed in Figure 5.7; the spreadsheet instructions are shown in Figure 5.8.

```
          A         B          C          D          E          F
 1
 2
 3              SOLAR RAYS ON TILTED SURFACE
 4
 5              ENTER PARAMETERS:
 6
 7              LATITUDE       35.00
 8              TILT           35.00
 9              DECLIN         18.00
10
11  ************************************************************
12
13              SIN(L-T)        0.00
14              COS(L-T)        1.00
15              SIN(D)          0.31
16              COS(D)          0.95
17
18                    HOUR     ANGLE
19                  800.00     61.61
20                  900.00     47.74
21                 1000.00     34.55
22                 1100.00     23.27
23                 1200.00     18.00
24                 1300.00     23.27
25                 1400.00     34.55
26                 1500.00     47.74
27                 1600.00     61.61
```

Figure 5.7: *Solar Rays on Tilted Surface Spreadsheet*

```
/GF$
B3:SOLAR RAY      C3:S ON TILT      D3:ED SURFAC      E3:E
B5:ENTER PAR      C5:AMETERS:
B7:LATITUDE
B8:TILT
B9:DECLIN
A11:/-*
/R:B11.F11
B13:SIN(L-T)
   C13:@SIN(@PI/180*(C7-C8))
B14:COS(L-T)
   C14:@COS(@PI/180*(C7-C8))
B15:SIN(D)
   C15:@SIN(@PI/180*C9)
B16:COS(D)
   C16:@COS(@PI/180*C9)
B18:/FR HOUR
   B19:800
   B20:100+B19
   /R:B21.B27:R
C18:/FR ANGLE
   C19:180/@PI*@ACOS(C13*C15+(C14*C16*@COS(@PI/180*
      (B19/100-12)*15)))
   /R:C20.C27:NNNNR
```

Figure 5.8: *Spreadsheet Instructions: Solar Rays on Tilted Surface*

6 Heating, Ventilating, and Air Conditioning

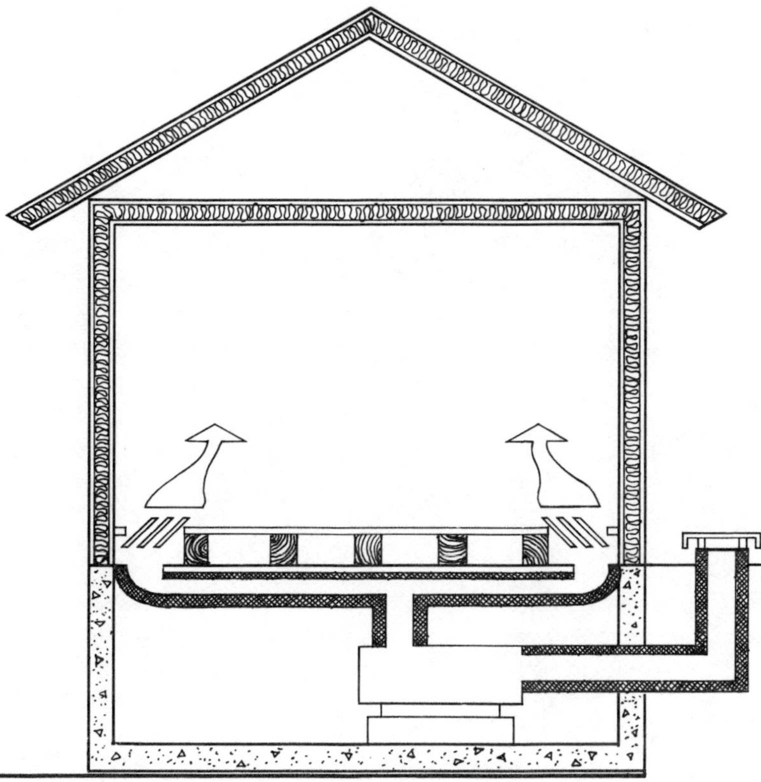

TEMPERATURE CONVERSION

PURPOSE

The spreadsheet in this example converts a series of user-specified temperatures from Fahrenheit to equivalent temperatures in degrees Centigrade, Kelvin, and Rankine. You can easily respecify this spreadsheet so the conversions are made from Centigrade or another scale.

DATA ENTRY

The spreadsheet calculates ten different conversions. You must specify the initial temperature and the temperature increment prior to calculating the converted temperatures.

EXAMPLE

Let's generate a series of temperature conversions from Fahrenheit to Centigrade, Kelvin and Rankine beginning with 20° F and continuing in 20° increments. The spreadsheet is displayed in Figure 6.1; the spreadsheet instructions appear in Figure 6.2.

	A	B	C	D	E	F
1						
2						
3		TEMPERATURE TABLES				
4						
5		ENTER PARAMETERS:				
6						
7		DEG F	20.00			
8		DEG IN	20.00			
9						
10	**					
11						
12		DEG F	DEG C	DEG K	DEG R	
13		20.00	-6.67	266.33	479.60	
14		40.00	4.44	277.44	499.60	
15		60.00	15.56	288.56	519.60	
16		80.00	26.67	299.67	539.60	
17		100.00	37.78	310.78	559.60	
18		120.00	48.89	321.89	579.60	
19		140.00	60.00	333.00	599.60	
20		160.00	71.11	344.11	619.60	
21		180.00	82.22	355.22	639.60	
22		200.00	93.33	366.33	659.60	

Figure 6.1: *Temperature Conversion Spreadsheet*

```
/GF$
B3:TEMPERATU     C3:RE TABLES
B5:ENTER PAR     C5:AMETERS:
B7:DEG F
B8:DEG INC
A10:/ - *
/R:B10.F10
B12:/FR
/R:C12.E12
B12:DEG F
   B13: +C7
   B14: +C8+B13
   /R:B15.B22:NR
C12:DEG C
   C13:5/9*(B13-32)
   /R:C14.C22:R
D12:DEG K
   D13:273+C13
   /R:D14.D22:R
E12:DEG R
   E13:459.6+B13
   /R:E14.E22:R
```

Figure 6.2: *Spreadsheet Instructions: Temperature Conversion*

BUILDING HEAT DESIGN

PURPOSE

We can easily use the VisiCalc program to calculate heat loss from buildings. The spreadsheet in this example uses U values (heat transfer coefficients) to calculate the building heat loss for various temperature differences. The basic equation which we use to calculate heat loss is:

$$\text{Loss} = T_{\text{diff}}UA$$

DATA ENTRY

Prior to using this spreadsheet you must enter the U values and areas for the walls, roof, floor, and windows of the building. The spreadsheet will then compute the heat loss for a range of inside-to-outside temperature differences of 10° F – 70° F.

	A	B	C	D	E	F
1						
2						
3		HEAT LOAD BUILDING				
4						
5		ENTER PARAMETERS:				
6						
7		WALL AREA	2000			
8		UWALL	.05			
9		ROOF AREA	3000			
10		UROOF	.08			
11		FLR AREA	3000			
12		UFLR	.07			
13		WIN AREA	1000			
14		UWIN	1.3			
15						
16	**********	**********	**********	**********	**********	**********
17						
18		TEM DIFF	LOSS			
19		10	18500			
20		20	37000			
21		30	55500			
22		40	74000			
23		50	92500			
24		60	111000			
25		70	129500			

Figure 6.3: *Building Heat Design Spreadsheet*

EXAMPLE

You wish to calculate the heat loss for a building. The walls have a total area of 2000 sq ft and a U value of .05; the roof, 3000 sq ft and .08; the floors, 3000 sq ft and .07; and the windows, 1000 sq ft and 1.3. The completed spreadsheet is displayed in Figure 6.3 (shown on the previous page); the instructions appear in Figure 6.4.

```
B3:HEAT LOAD      C3:" BUILDING
B5:ENTER PAR      C5:AMETERS:
B7:WALL AREA
B8:UWALL
B9:ROOF AREA
B10:UROOF
B11:FLR AREA
B12:UFLR
B13:WIN AREA
B14:UWIN
A16:/-*
/R:B16.F16
B18:/FR TEM DIFF
   B19:10
   B20:10+B19
   /R:B21.B25:R
C18:/FR LOSS
   C19:+B19*((C7*C8)+(C9*C10)+(C11*C12)+(C13*C14))
   /R:C20.C25:RNN NNN NNN
```

Figure 6.4: *Spreadsheet Instructions: Building Heat Design*

PIPE FLOW

PURPOSE

This spreadsheet calculates the pressure drop per meter in a circular pipe. The equation we use is based on laminar flow for which the Reynolds number must be less than 2000. We use the following equations:

$$Re = \frac{\text{dens} \times \text{vel} \times \text{dia}}{\text{viscosity}}$$

$$\text{Press} = \frac{64\text{vel}^2 \times \text{dens}}{2/Re}$$

DATA ENTRY

To generate the spreadsheet, you must specify several parameters, including the fluid's velocity, viscosity, and density. Since the pressure drop per meter is computed for several diameters of pipe, you must also enter the starting diameter and the diameter increment.

EXAMPLE

Let's compute the pressure drop and Reynolds number for a fluid flowing in a circular pipe. All quantities are taken as metric. They are: a velocity of .02, a density of 1, and a viscosity of 2.07E−5. We will use a series of pipe diameters from .2 to 2 meters. The spreadsheet is shown in Figure 6.5; the instructions for the spreadsheet appear in Figure 6.6.

```
        A         B         C         D         E         F

 1
 2
 3          PIPE FLOW
 4
 5          ENTER PARAMETERS:
 6
 7          VELOCITY      0.02
 8          DENSITY       1.00
 9          VISCOSITY .0000207
10          ST. DIA.      0.20
11          DIA. INC.     0.20
12
13 ************************************************************
14
15          PRESSURE DROP PER METER
16
17              DIAM.    RE NO.    DEL P
18               0.20    193.24 6.624E-5
19               0.40    386.47 3.312E-5
20               0.60    579.71 2.208E-5
21               0.80    772.95 1.656E-5
22               1.00    966.18 1.325E-5
23               1.20   1159.42 1.104E-5
24               1.40   1352.66 9.463E-6
25               1.60   1545.89  8.28E-6
26               1.80   1739.13  7.36E-6
27               2.00   1932.37 6.624E-6
```

Figure 6.5: *Pipe Flow Spreadsheet*

```
/GF$
B3:PIPE FLOW
B5:ENTER PAR      C5:AMETERS:
B7:VELOCITY
B8:DENSITY
B9:VISCOSITY
C9:/FG
B10:ST. DIA.
B11:DIA. INC.
A13:/- *
/R:B13.F13
B15:PRESSURE      C15:DROP PER      D15:METER
B17:/FR
```

Figure 6.6: *Spreadsheet Instructions: Pipe Flow*

```
/R:C17.D17
B17:DIAM.
   B18:+C10
   B19:+B18+C11
   /R:B20.B27:RN
C17:RE NO.
   C18:+C7*C8/C9*B18
   /R:C19.C27:NNNR
D17:DEL P
   D18:/FG 64/C18*C8*C7*C7/2
   /R:D19.D27:RNNN
```

Figure 6.6, Continued

INSULATED PIPE

PURPOSE

The spreadsheet in this example computes the heat loss for an insulated cylindrical pipe for varying pipe lengths and temperature differences. Heat loss per foot is computed from the equation:

$$q = \frac{2\pi}{R_1 + R_2 + R_3 + R_4}$$

where R_1, R_2, etc., are the individual heat transfer coefficients.

DATA ENTRY

This spreadsheet requires the entry of a series of parameters before the heat transfer calculations can be performed. You must enter the convective and conductive heat transfer coefficients for both the inside and outside of the pipe. Additionally, you enter the inner and outer diameters of the pipe and the outer diameter of the insulation.

EXAMPLE

You wish to use the VisiCalc program to calculate the heat loss for an insulated pipe. The inner and outer diameters of the pipe are 3 and 4 inches, respectively. The outer diameter of the insulation is 7 inches. The inside and outside convective coefficients are 950 and 5, respectively. The inside and outside conductive coefficients are 25 and .1, respectively. The program calculates the four individual heat transfer coefficients, and the ratio of heat loss to length (*Q*/*L*). It then generates a table of values for heat loss for pipes from 10 to 100 feet, over a range of temperature differences from 10 to 50 degrees. The spreadsheet and its instructions are shown in Figures 6.7 and 6.8, respectively.

```
          A          B          C          D          E          F          G
 1
 2
 3              HEAT TRANSFER-INSULATED PIPE
 4
 5              ENTER PARAMETERS:
 6
 7              HC(I)         950.00
 8              HC(O)           5.00
 9              K(I)           25.00
10              K(O)            0.10
11              D1              3.00
12              D2              4.00
13              D3              7.00
14
15   ******************************************************
16
17              R1         5.848E-5
18              R2         .0115073
19              R3         5.596158
20              R4         .0047619
21
22              Q/L             1.12
23
24                                 TEMP DIFF DEG F
25              LENGTH FT      10.00      20.00      30.00      40.00      50.00
26                  10.00     111.95     223.90     335.85     447.80     559.75
27                  20.00     223.90     447.80     671.70     895.60    1119.50
28                  30.00     335.85     671.70    1007.55    1343.40    1679.25
29                  40.00     447.80     895.60    1343.40    1791.20    2239.00
30                  50.00     559.75    1119.50    1679.25    2239.00    2798.75
31                  60.00     671.70    1343.40    2015.10    2686.80    3358.50
32                  70.00     783.65    1567.30    2350.95    3134.60    3918.25
33                  80.00     895.60    1791.20    2686.80    3582.40    4478.01
34                  90.00    1007.55    2015.10    3022.65    4030.20    5037.76
35                 100.00    1119.50    2239.00    3358.50    4478.01    5597.51
```

Figure 6.7: *Insulated Pipe Spreadsheet*

```
/GF$
B3:HEAT TRAN     C3:SFER – INSU     D3:LATED PIP     E3:E
B5:ENTER PAR     C5:AMETERS:
B7:HC(I)
B8:HC(O)
B9:K(I)
B10:K(O)
B11:D1
B12:D2
B13:D3
A15:/ – *
/R:B15.G15
B17:R1
   C17:/FG 2/C7/C11/12
B18:R2
   C18:/FG @LN(C12/C11)/C9
B19:R3
   C19:/FG @LN(C13/C12)/C10
B20:R4
   C20:/FG 2/C8/C13/12
B22:Q/L
   C22:2*@PI/@SUM(C17.C20)
B25:LENGTH FT
   B26:10
   B27:10+B26
   /R:B28.B35:R
D24:TEMP DIFF     E24:" DEG F
   C25:10
   D25:10+C25
   /R:E25.G25:R
C26: +C22*B26*C25
/R:D26.G26:NNR
/R.G26:C27.C35:NRN NRN NRN NRN NRN
```

Figure 6.8: *Spreadsheet Instructions: Insulated Pipe*

INSULATION ANALYSIS

PURPOSE

The spreadsheet in this example is designed to evaluate the economic effect of adding insulation to a building. We use the cost of the additional insulation, the projected fuel savings, and the interest rate on investments to calculate the net present value of the insulation. We have set up the spreadsheet for a ten-year cycle; you may modify this as desired for your own needs.

```
          A          B          C          D          E          F
 1
 2
 3              INSULATION ANALYSIS
 4
 5              ENTER PARAMETERS:
 6
 7              INIT FUEL SAVINGS       500.00
 8              FUEL ESC RATE             8.00
 9              INSULATION COST        5000.00
10              INTEREST RATE            10.00
11
12 ************************************************************
13
14                     YEAR     SAVED
15                     1.00    500.00
16                     2.00    540.00
17                     3.00    583.20
18                     4.00    629.86
19                     5.00    680.24
20                     6.00    734.66
21                     7.00    793.44
22                     8.00    856.91
23                     9.00    925.47
24                    10.00    999.50
25
26              NPV           2243.28
```

Figure 6.9: *Insulation Analysis Spreadsheet*

DATA ENTRY

You will enter the cost of installing the new insulation, the first-year fuel savings, the projected fuel cost escalation rate, and the rate at which invested funds earn money.

EXAMPLE

You are evaluating the feasibility of adding insulation to an office building. The new insulation will cost $5000. You project that the first-year fuel savings will be $500. Fuel is expected to increase in cost at 8% per year. Invested funds earn interest at 10% per annum. The spreadsheet for this example appears in Figure 6.9 (shown on opposite page); and the spreadsheet instructions in Figure 6.10.

```
/GF$
B3:INSULATIO      C3:N ANALYSI      D3:S
B5:ENTER PAR      C5:AMETERS:
B7:INIT FUEL      C7:" SAVINGS
B8:FUEL ESC       C8:RATE
B9:INSULATIO      C9:N COST
B10:INTEREST      C10:RATE
A12:/- *
/R:B12.F12
B14:/FR YEAR
   B15:1
   B16:1+B15
   /R:B17.B24:R
C14:/FR SAVED
   C15:+D7
   C16:+C15*(1+(D8/100))
   /R:C17.C24:RN
B26:NPV
   C26: -D9+@NPV(D11/100,C15.24)
```

Figure 6.10: *Spreadsheet Instructions: Insulation Analysis*

7

Control Systems

FREQUENCY RESPONSE

PURPOSE

This spreadsheet allows you to compute the frequency response for a transfer function of the form:

$$T(s) = \frac{K_1(\tau_2 s + 1)}{s^{N_3}(\tau_4 s + 1)(\tau_5 s + 1)\left(-\frac{s^2}{\omega_7^2} + \frac{2\delta_6 s}{\omega_7} + 1\right)}$$

The table produces the frequency in hertz and radians/second as well as the phase angle, magnitude, and magnitude in decibels.

DATA ENTRY

First enter the values for K_1 and the numerator time constant, τ_2. Then enter N_3, the order of the pole at the origin. Next, enter the two time constants, τ_4 and τ_5. Then enter the values for the damping coefficient δ_6 and the natural frequency ω_7. Finally, you specify the initial frequency. The design table presents ten frequencies in multiples of the initial frequency: $\times$ 1, $\times$ 2, $\times$ 5, $\times$ 10, $\times$ 20, $\times$ 50, $\times$ 100, $\times$ 500, and $\times$ 1000. The phase angle and magnitudes are computed at each frequency. You should note that column E is used and that column A contains intermediate calculations from A50 through A179, not shown in the spreadsheet.

EXAMPLE

You want to determine the frequency response for a transfer function having the following values: $K_1 = 4$, $\tau_2 = 0.5$ sec, $N_3 = 1$, $\tau_4 = 2$ sec, $\tau_5 = 0$ sec, $\delta_6 = 0.2$, and $\omega_7 = 8$ rad/sec. You specify the initial frequency as 0.01592 Hz. If you do not wish to use the quadratic term, set δ_6 to 0, and make ω_7 very large compared to frequencies of interest. The results are displayed in Figure 7.1 and the spreadsheet instructions are given in Figure 7.2.

```
         A          B          C          D          E          F
 1                TRANSFER FUNCTION
 2                FREQUENCY RESPONSE
 3
 4                ENTER PARAMETERS:
 5
 6                     K1  =      4.00
 7                     T2  =      0.50 SEC
 8                     N3  =      1.00
 9                     T4  =      2.00 SEC
10                     T5  =      0.00 SEC
11                     D6  =      0.20
12                     W7  =      8.00 RAD/SEC
13                     IN FREQ =    0.01592 HZ
14
15 ************************************************************
16
17      FREQ   FRQ R/S        ANG        MAG         DB
18
19      0.02      0.10    -98.74      39.27      31.88
20      0.03      0.20   -106.67      18.67      25.42
21      0.08      0.50   -122.41       5.85      15.34
22      0.16      1.00   -129.78       2.03       6.14
23      0.32      2.00   -127.05       0.73      -2.77
24      0.80      5.00   -128.41       0.33      -9.75
25      1.59     10.00   -236.85       0.14     -17.38
26      3.18     20.01   -263.50       0.01     -40.55
27      7.96     50.01   -267.96       0.00     -65.61
28     15.92    100.03   -269.01       0.00     -83.83
```

Figure 7.1: *Frequency Response Spreadsheet*

```
/GF$
B1:TRANSFER        C1:FUNCTION
B2:FREQUENCY       C2:" RESPONSE
B4:ENTER PAR       C4:AMETERS:
B6:/FR K1 =
B7:/FR T2 =        D7:" SEC
B8:/FR N3 =
B9:/FR T4 =        D9:" SEC
B10:/FR T5 =       D10:" SEC
B11:/FR D6 =
B12:/FR W7 =       D12:" RAD/SEC
B13:IN FREQ =      C13: /FG            D13:" HZ
A15:/ - *
/R:B15.F15
```

Figure 7.2: *Spreadsheet Instructions: Frequency Response*

```
A17:/FR FREQ
   A19: +C13
   A20:2*C13
   A21:5*C13
   A22:10*A19
   /R:A23.A24:R
   A25:100*A19
   /R:A26.A27:R
   A28:1000*A19
B17:/FR FRQ R/S
   B19:2*@PI*A19
   /R:B20.B28:R
C17:/FR ANG
   C19:180*(B50+B70-B90-B110)/@PI
   /R:C20.C28:RRRR
D17:/FR MAG
   D19: +B130/(B150*B170)
   /R:D20.D28:RRR
E17:/FR DB
   E19:20*@LOG10(D19)
   /R:E20.E28:R
;COMMENT INTERMEDIATE VARIABLES FOLLOW
B50:.5*@PI*((@ABS(C6)-C6)/C6
/R:B51.B59:NNN
B70:@ATAN(C7*B19)-(.5*@PI*C8)
/R:B71.B79:NRN
B90:@ATAN(C9*B19)+@ATAN(C10*B19)
/R:B91.B99:NRNR
B110:(@ATAN((B19+(C12*@SQRT(1-(C11*C11)))/
   (C11*C12)))+@ATAN((B19-(C12*@SQRT(1-(C11*C11)))/(C11*C12)))
/R:B111.B119:RNNNNNRNNNNN
B130:@SQRT((C6*C6)*(1+((C7*B19)^2)))
/R:B131.B139:NNNR
B150:@SQRT(((B19*B19)^C8)*(1+((B19*C9)^2)*(1+((B19*C10)^2)))
/R:B151.B159:RRNRNRN
B170:@SQRT(((1-((B19/C12)^2)^2)+((2*C11*B19/C12)^2))
/R:B171.B179:RNNRN
```

Figure 7.2, Continued

PID DIGITAL CONTROL

PURPOSE

This spreadsheet permits you to design a digital PID control algorithm, given the parameters of a first-order process model. Initial PI coefficients, PID digital control algorithm coefficients, and the algorithm itself are tabulated in the spreadsheet. The design equations are:

$$\text{Proportional Gain} = P_K = \frac{1}{K\,[\exp(T/T_c) - 1]}$$

$$\text{Integral Time} = T_i = \frac{T}{\exp(T/T_c) - 1}$$

$$A_0 = P_K\left(1 + \frac{T}{T_i} + \frac{T_d}{T}\right)$$

$$A_1 = -P_K\left(1 + 2\frac{T_d}{T}\right)$$

$$A_2 = P_K\left(\frac{T_d}{T}\right)$$

where:

K = process gain,
T_c = process lag time-constant,
T = sampling time, and
T_d = derivative time.

Note: PID denotes Proportional, Integral, and Derivative.

DATA ENTRY

You first enter the process gain, *K*, and the lag time-constant. Next enter the system sampling time. At this point, the program tabulates initial PI coefficients for you as guidance for design. You can then enter final PID coefficients as parameters to yield the digital PID control algorithm coefficients A_0, A_1, and A_2, as well as the actual algorithm. Note that the spreadsheet continues to line 28.

EXAMPLE

Let's design a controller with the following process parameters: proportional gain = 0.01, lag time-constant = 21, and sampling time = 1. The initial proportional gain is found to be 2050.4 and integral time to be 20.5. You then decide to use a proportional gain of 2000, an integral time constant of 20, and a derivative time constant of 0.1. The spreadsheet in Figure 7.3 presents the A_0, A_1, and A_2 coefficients and the digital control algorithm. Instructions for this spreadsheet are displayed in Figure 7.4.

```
         A       B        C          D         E         F
 1          PID DIGITAL CONTROL
 2
 3          ENTER PARAMETERS:
 4
 5          GAIN    K=      0.01
 6          LAG T.C.=      21.00
 7          SAMPLE T=       1.00
 8
 9          PRO GAIN=   2000.00
10          INT.T.C.      20.00
11          DER.T.C.=      0.10
12
13  ******************************************************************
14
15                 PK=  2050.40
16                 IT=    20.50
17
18                 A0=  2300.00
19                 A1= -2400.00
20                 A2=      200
21
22                   DIGITAL   ALGORITHM
23
24       M(K) = M(K-1) +    2300.00*E(K) +
25
26                         -2400.00*E(K-1) +
27
28                              200*E(K-2)
```

Figure 7.3: *PID Digital Control Spreadsheet*

```
/GF$
B1:PID DIGIT        C1:AL CONTRO      D1:L
B3:ENTER PAR        C3:AMETERS:
B5:GAIN    K=
B6:LAG T.C. =
B7:SAMPLE T=
B9:PRO GAIN=
B10:INT.T.C. =
B11:DER.T.C. =
A13:/- *
/R:B13.F13
B15:/FR PK=
   C15: +1/(C5*(@EXP(C7/C6)-1))
B16:/FR IT=
   C16: +C7/(@EXP(C7/C6)-1)
B18:/FR A0=
   C18: +C9*(1+(C7/C10)+(C11/C7))
B19:/FR A1=
   C19: -C9*(1+(2*C11/C7))
B20:/FR A2=
   C20: +C9*C11/C7
B22:DIGITAL         C22:ALGORITHM
A24:"  M(K) =       B24:M(K-1) +
C24: +C18           D24:"*E(K) +
C26: +C19           D26:"*E(K-1) +
C28: +C20           D28:"*E(K-2)
```

Figure 7.4: *Spreadsheet Instructions: PID Digital Control*

PID ANALOG CONTROL

PURPOSE

We use this spreadsheet to compute the PID analog control coefficients for the Ziegler-Nichols, Cohen-Coon, and 3C tuning methods. Specifying the process parameters—gain, lag time, and delay time—results in a table of P, PI, and PID coefficients for each of the three tuning methods. The process parameters are defined in the process reaction curve step response shown below:

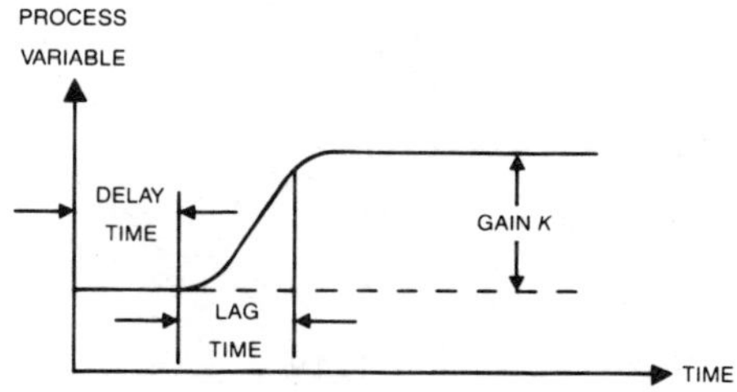

Note: PID denotes Proportional, Integral, and Derivative.

DATA ENTRY

After we obtain the process parameters, we enter the process gain, K, the lag time, and the delay time as design parameters into the spreadsheet. The values for the P, PI, and PID coefficients are produced in the design table, which extends through line 26 of the spreadsheet.

EXAMPLE

We wish to obtain the PID coefficients for a process having a gain of 1.0, a lag time of 1.39 sec, and a delay time of 1.19 sec. The spreadsheet is shown in Figure 7.5. Instructions for this spreadsheet are shown in Figure 7.6. Using the standard form for the PID transfer function:

$$T(s) = P\left(1 + \frac{1}{Is} + Ds\right)$$

where:

P = proportional gain,
I = integral time, and
D = derivative time.

with the 3C method we have

$$T(s) = 1.59(1 + \frac{1}{0.92s} + 0.44s)$$

as the PID controller transfer function.

```
           A          B          C          D
 1                PID ANALOG CONTROL
 2
 3                ENTER PARAMETERS:
 4
 5                GAIN K =         1.00
 6
 7                LAG T =          1.39
 8
 9                DELAY T =        1.19
10
11 ************************************
12
13                     P          I          D
14 ------------------------------------
15 ZIEGLER-         1.17
16                  1.05       3.96
17 NICHOLS          1.40       2.38       0.60
18 ------------------------------------
19 COHEN            1.50
20                  1.13       1.48
21 COON             1.85       2.30       0.38
22 ------------------------------------
23                  1.40
24 3C               1.07       1.18
25                  1.59       0.92       0.44
```

Figure 7.5: *PID Analog Control Spreadsheet*

```
/GF$
B1:PID ANALO      C1:G CONTROL
B3:ENTER PAR      C3:AMETERS:
B5:GAIN K =
```

Figure 7.6: *Spreadsheet Instructions: PID Analog Control*

```
B7:/FR LAG T =
B9:DELAY T=
A11:/- *
/R:B11.D11
B13:/FR P          C13:/FR I          D13:/FR D
A14:/- -
/R:B14.D14
A15:ZIEGLER-       A17:NICHOLS
   B15: +C7/(C9*C5)
   B16:.9*B15
   B17:1.2*B15
   C16:3.33*C9
   C17:2*C9
   D17:.5*C9
A18:/- -
/R:B18.D18
A19:COHEN-         A21:COON
   B19:(C7/C9+.333)/C5
   B20:(.9*C7/C9+.082)/C5
   B21:(1.35*C7/C9+.27)/C5
   C20:3.33*C9*(1+(C9/C7/11)/(1+(2.2*C9/C7))
   C21:2.5*C9*(1+(C9/(C7*5)))/(1+(.6*C9/C7))
   D21:.37*C9/(1+(.2*C9/C7))
A22:/- -
/R:B22.D22
A24:"  3C
   B23:1.208*((C7/C9)^.956)/C5
   B24:.928*((C7/C9)^.946)/C5
   B25:1.37*((C7/C9)^.95)/C5
   C24:.928*C7*((C9/C7)^.583)
   C25:.74*((C9/C7)^.738)*C7
   D25:.365*C7*((C9/C7)^.95)
```

Figure 7.6, Continued

MOTOR TORQUE—DIRECT DRIVE

PURPOSE

Given the parameters of a directly-driven cylindrical load, the load weight, inertia, and torque are calculated for a given change in RPM over a specified time interval. This spreadsheet thus provides information that allows you to size and specify a motor application. The design equations are as follows:

$$\text{Weight} = W = \pi L \rho (R_2^2 - R_1^2)$$

$$\text{Inertia} = I_L = W(R_2^2 - R_1^2)/2$$

$$\text{Torque} = \frac{2\pi}{60.386}(I_L + I_M)\frac{\Delta\omega}{\Delta t}$$

where:

ρ = material density
I_M = motor inertia
$\Delta\omega$ = change in RPM
Δt = change in time

and R_1, R_2 and L are defined in the diagram below:

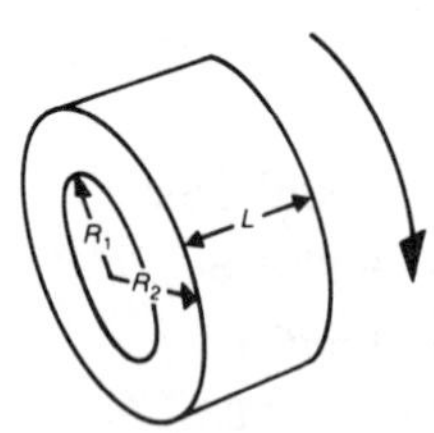

DATA ENTRY

You first specify the load's radii, length, and material density. Next, you enter the change in RPM and the period over which it is to occur. Finally, you enter the inertia of the motor. The spreadsheet gives you the load weight, load inertia, and load torque.

EXAMPLE

You have a solid aluminum cylinder. Since there is no inner radius in this case, R_1 is equal to zero. R_2 is equal to 2 in. and the length is equal to 4 in. For aluminum, the density is 1.54 oz/in^3. You want to accelerate the load to 100 rpm in 1.0 second, and the motor inertia is given as 10 oz-in.^{2}. The resulting spreadsheet is shown in Figure 7.7. Instructions for this spreadsheet are given in Figure 7.8.

```
        A         B            C          D
 1            MOTOR TORQUE-
 2            DIRECT   DRIVE
 3
 4            ENTER PARAMETERS:
 5
 6            RADIUS 1=        0.00 IN
 7            RADIUS 2=        2.00 IN
 8            LENGTH   =       4.00 IN
 9            DENSITY  =       1.54 OZ/CU IN
10            DEL. RPM=      100.00 RPM
11            DEL.TIME=        1.00 SEC
12            INERT(M)=       10.00 OZ-SQ IN
13
14   ************************************
15
16            WEIGHT          77.41 OZ
17
18            INERT (L)      154.82 OZ-SQ IN
19
20            TORQUE           4.47 OZ-IN
```

Figure 7.7: *Motor Torque—Direct Drive Spreadsheet*

```
/GF$
B1:MOTOR TOR      C1:QUE –
B2:DIRECT         C2:DRIVE
B4:ENTER PAR      C4:AMETERS:
B6:RADIUS 1 =     D6:" IN
B7:RADIUS 2 =     D7:" IN
B8:LENGTH  =      D8:" IN
B9:DENSITY  =     D9:" OZ/CU IN
B10:DEL. RPM =    D10: " RPM
```

Figure 7.8: *Spreadsheet Instructions: Motor Torque—Direct Drive*

```
B11:DEL.TIME =      D11:" SEC
B12:INERT(M) =      D12:" OZ - SQ IN
A14:/ - *
/R:B14.D14
B16:WEIGHT          C16:@PI*C8*C9*((C7*C7)+(C6*C6))  D16:" OZ
B18:INERT(L)        C18:.5*C16*((C7*C7)+(C6*C6))     D18:" OZ - SQ IN
B20:TORQUE          C20:2*@PI*(C18+C12)*C10/(60*386*C11)
D20:" OZ - IN
```

Figure 7.8, Continued

MOTOR TORQUE—LEADSCREW DRIVE

PURPOSE

Given the parameters of a leadscrew-driven load, and a change in velocity over a specified time, this spreadsheet calculates the frictional force, the load inertia, the leadscrew inertia, the frictional torque, the acceleration torque, and the total torque. This spreadsheet allows you to size and specify a motor application. The design equations are shown below and illustrated in the accompanying diagram.

$$\text{Frictional Force} = F = \mu W$$

$$\text{Load Inertia} = I_L = \frac{W}{(2\pi P)^2}$$

$$\text{Leadscrew Inertia} = I_S = \frac{\pi L \rho (D/2)^4}{2}$$

$$\text{Frictional Torque} = T_F = \frac{F}{2\pi Pe}$$

$$\text{Acceleration Torque} = T_A = \frac{2\pi P\Delta V}{386e\Delta t}[I_L + I_S + I_M]$$

$$\text{Total Torque} = T_F + T_A$$

where:

W = load weight,
μ = coefficient of static friction,
P = pitch,
e = leadscrew efficiency,
D = leadscrew diameter,
L = leadscrew length,
ρ = leadscrew material density,
ΔV = change in velocity,
Δt = change in time, and
I_M = motor inertia.

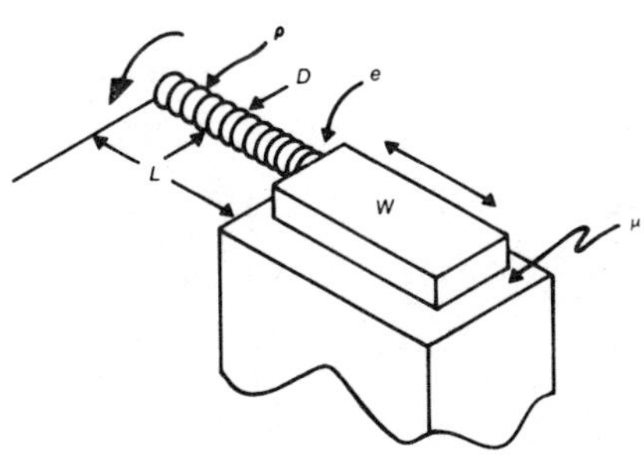

DATA ENTRY

You first specify the load weight and coefficient of static friction. Next enter the screw pitch, efficiency, diameter, length and material density. Then enter the change in linear velocity, and the time interval over which the change is to occur. Finally, enter the inertia of the motor. The spreadsheet gives you the frictional force, the inertia of the load and leadscrew, and the frictional, acceleration, and total torques.

EXAMPLE

You have a 100 lb load sliding on a table with a coefficient of static friction of 0.15. A steel leadscrew has a pitch of 5 threads/in., an efficiency of 90%, a diameter of 1 in., a length of 36 in., and a material density of 4.48 oz/in.^{3}. You wish to accelerate the load to 1 in./sec in 0.1 second, and the motor inertia is given as 5 oz -in.^{2}. The spreadsheet results are displayed in Figure 7.9. Instructions for this spreadsheet are given in Figure 7.10.

```
          A         B         C         D
 1                MOTOR TORQUE-
 2                LEADSCREW DRIVE
 3
 4                ENTER PARAMETERS:
 5
 6                LOAD WT       100.00 LBS
 7                CO ST FR        0.15
 8                PITCH(S)        5.00 TRDS/IN
 9                EFFIC(S)       90.00 %
10                DIA(S)          1.00 IN
11                LENG(S)        36.00 IN
12                DENS(S)         4.48 OZ/CU IN
13                DEL VEL         1.00 IN/SEC
14                DEL TIME        0.10 SEC
15                INERT (M)       5.00 OZ-SQ IN
16
17   ***********************************
18
19                FRI FOR       240.00 OZ
20
21                INERT(L)        1.62 OZ-SQ IN
22
23                INERT(S)       15.83 OZ-SQ IN
24
25                FRI TORQ        8.49 OZ IN
26
27                ACC TORQ       20.31 OZ IN
28
29                TOT TORQ       28.79 OZ IN
```

Figure 7.9: *Motor Torque—Leadscrew Drive*

```
/GF$
B1:MOTOR TOR     C1:QUE–
B2:LEADSCREW     C2:" DRIVE
B4:ENTER PAR     C4:AMETERS:
```

Figure 7.10: *Spreadsheet Instructions: Motor Torque—Leadscrew Drive*

```
B6:LOAD WT        D6:" LBS
B7:CO ST FR
B8:PITCH(S)       D8:" TRDS/IN
B9:EFFIC(S)       D9:" %
B10:DIA(S)        D10:" IN
B11:LENG(S)       D11:" IN
B12:DENS(S)       D12:" OZ/CU IN
B13:DEL VEL       D13:" IN/SEC
B14:DEL TIME      D14:" SEC
B15:INERT(M)      D15:" OZ-SQ IN
A17:/- *
/R:B17.D17
B19:FRI FOR
   C19:16*C7*C6
   D19:" OZ
B21:INERT(L)
   C21:16*C6/((2*@PI*C8)^2)
   D21:" OZ- SQ IN
B23:INERT(S)
   C23:.5*@PI*C11*C12*((C10/2)^4)
   D23:" OZ-SQ IN
B25:FRI TORQ
   C25:100*C19/(2*@PI*C8*C9)
   D25:" OZ IN
B27:ACC TORQ
   C27:2*@PI*C8*C13*(C15+C21+C23)/(3.86*C14*C9)
   D27:" OZ IN
B29:TOT TORQ
   C29:+C25+C27
   D29:" OZ IN
```

Figure 7.10, Continued

MOTOR TORQUE—TANGENTIAL DRIVE

PURPOSE

Given the parameters of a tangential-driven load, values for the load, pulley, and belt inertias, and the frictional, acceleration, and total torques are calculated for a given change in velocity over a specified time interval. This spreadsheet presents information that allows you to size and specify a motor application. The design equations are shown below and illustrated in the accompanying diagram.

$$\text{Load Inertia} = I_L = W_L R^2$$

$$\text{Pulley Inertia} = I_P = \frac{W_P R^2}{2} N$$

$$\text{Belt inertia} = I_B = W_B R^2$$

$$\text{Frictional Torque} = T_F = FR$$

$$\text{Acceleration Torque} = T_A = \frac{\Delta V}{\Delta t} \frac{I_L + I_P + I_B + I_M}{386R}$$

$$\text{Total Torque} = T_F + T_A$$

where:

- W_L = Load weight,
- R = pulley radius,
- N = number of pulleys,
- W_P = pulley weight,
- W_B = belt weight,
- F = Force due to static friction,
- I_M = Motor inertia,
- ΔV = Change in velocity, and
- Δt = Change in time.

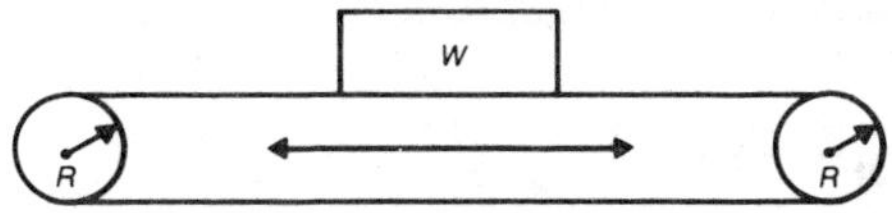

DATA ENTRY

You first enter the load weight, the pulley radius, and the number of pulleys. Next enter the pulley and belt weights, the frictional force, and the motor inertia. Finally, you specify the change in velocity and the change in time over which it is to occur. The spreadsheet displays the load, pulley, and belt inertias, and the frictional, acceleration, and total torques.

EXAMPLE

You have a 10 lb load driven by a two-pulley system, with a pulley radius of 2 in. The pulley weight is 5 oz and the belt weight is 15 oz. You measure a static friction of 25 oz, and the motor inertia is 2 oz-in^2. You wish to accelerate the load to 30 in./sec in 0.1 second. The spreadsheet results are shown in Figure 7.11. Instructions for this spreadsheet are given in Figure 7.12.

	A	B	C	D
1		MOTOR TORQUE		
2		TANGENTIAL DRIVE		
3				
4		ENTER PARAMETERS:		
5				
6		WT (L)	10.00	LB
7		RADIUS	2.00	IN
8		PULLEY#	2.00	
9		WT (P)	5.00	OZ
10		WT (B)	15.00	OZ
11		FRIC(ST)	25.00	OZ
12		INERT(M)	2.00	OZ-SQ IN
13		DEL VEL	30.00	IN/SEC
14		DEL TIME	0.10	SEC
15				
16	*************************************			
17				
18		INERT(L)	640.00	OZ SQ IN
19		INERT(P)	20.00	OZ SQ IN
20		INERT(B)	60.00	OZ SQ IN
21		FRI TORQ	50.00	OZ IN
22		ACC TORQ	280.57	OZ IN
23		TOT TORQ	330.57	OZ IN

Figure 7.11: *Motor Torque—Tangential Drive Spreadsheet*

```
/GF$
B1:MOTOR TOR     C1:QUE
B2:TANGENTIA     C2:L DRIVE
B4:ENTER PAR     C4:AMETERS:
B6:WT (L)        D6:" LB
B7:RADIUS        D7:" IN
B8:PULLEY#
B9:WT (P)        D9:" OZ
B10:WT (B)       D10:" OZ
B11:FRIC(ST)     D11:" OZ
B12:INERT(M)     D12:" OZ-SQ IN
B13:DEL VEL      D13:" IN/SEC
B14:DEL TIME     D14:" SEC
A16:/-*
/R:B16.F16
B18:INERT(L)
   C18:16*C6*C7*C7
   D18:" OZ SQ IN
B19:INERT(P)
   C19:.5*C9*C7*C7*C8
   D19:" OZ SQ IN
B20:INERT(B)
   C20:+C10*C7*C7
   D20:" OZ SQ IN
B21:FRI TORQ
   C21:+C11*C7
   D21:" OZ IN
B22:ACC TORQ
   C22:+C13*(C18+C19+C20+C12)/(386*C7*C14)
   D22:" OZ IN
B23:TOT TORQ
   C23:+C21+C22
   D23:" OZ IN
```

Figure 7.12: *Spreadsheet Instructions: Motor Torque—Tangential Drive*

MOTOR TORQUE—GEAR DRIVE

PURPOSE

Given the parameters of a gear-driven cylindrical load, the load, gear 1, and gear 2 inertias and the total torque are calculated for a given change in velocity over a specified time interval. This spreadsheet presents information that allows you to size and specify a motor application. The design equations are shown below and illustrated in the accompanying diagram.

$$\text{Load Inertia} = I_L = \frac{W_L}{2} R^2 \left(\frac{T_2}{T_1}\right)^2$$

$$\text{Gear 1 Inertia} = I_1 = \frac{W_1}{2} R_1^2 \left(\frac{T_2}{T_1}\right)^2$$

$$\text{Gear 2 Inertia} = I_2 = \frac{W_2}{2} R_2^2$$

$$\text{Total Torque} = \frac{2\pi \Delta \omega}{60 \times 386 \Delta t} (I_L + I_1 + I_2 + I_M)$$

where:

W_L = cylindrical load weight,
R = cylindrical load radius,
W_1 = gear 1 weight,
R_1 = gear 1 radius,
T_1 = gear 1 number of teeth,
W_2 = gear 2 weight,
R_2 = gear 2 radius,
T_2 = gear 2 number of teeth,
$\Delta \omega$ = change in RPM, and
Δt = change in time.

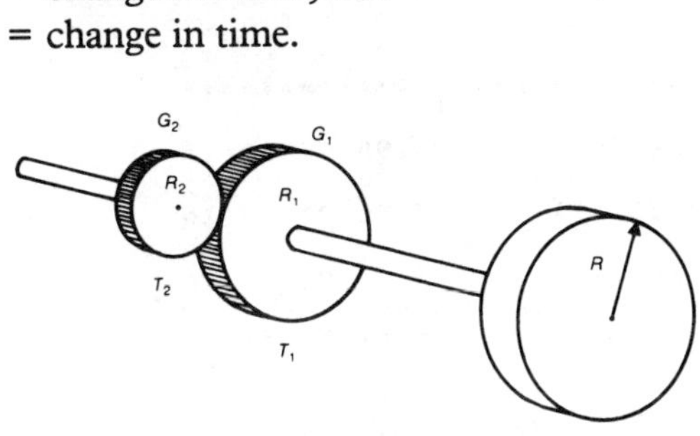

DATA ENTRY

You first enter the load weight and radius. Next enter the gear 1 weight, radius, and number of teeth. Then enter the gear 2 weight, radius, and number of teeth. Finally, you specify the motor inertia, the change in velocity, and change in time over which it is to occur. The spreadsheet displays the load, gear 1, and gear 2 inertias, as well as the total torque.

EXAMPLE

You have a 174.2-oz load with a radius of 3 in. The load gear weighs 7.04 oz, has a radius of 3 in. and has 20 teeth. The drive gear weighs 3.5 oz, has a radius of 1.5 in., and has 10 teeth. You know that the motor inertia is 10 oz-in^2, and you wish to accelerate the load by 100 RPM in 1 second. The spreadsheet results are shown in Figure 7.13. Instructions for this spreadsheet are given in Figure 7.14.

	A	B	C	D
1		MOTOR TORQUE		
2		GEAR DRIVE		
3				
4		ENTER PARAMETERS:		
5				
6		WT (LO)	174.20	OZ
7		RAD (L)	3.00	IN
8		WT (G1)	7.04	OZ
9		RAD(G1)	3.00	IN
10		#TEETH 1	20.00	
11		WT (G2)	3.50	OZ
12		RAD (G2)	1.50	IN
13		#TEETH 2	10.00	
14		INERT(M)	10.00	OZ-SQ IN
15		DEL RPM	100.00	
16		DEL TIME	1.00	SEC
17				
18	**********	**********	**********	**********
19				
20		INER(L)	195.98	OZ SQ IN
21				
22		INER(G1)	7.92	OZ SQ IN
23				
24		INER(G2)	3.94	OZ SQ IN
25				
26		TORQUE	5.91	OZ IN

Figure 7.13: *Motor Torque—Gear Drive Spreadsheet*

```
/GF$
B1:MOTOR TOR      C1:QUE
B2:GEAR DRIV      C2:E
B4:ENTER PAR      C4:AMETERS
B6:WT (LO)        D6:" OZ
B7:RAD (L)        D7:" IN
B8:WT (G1)        D8:" OZ
B9:RAD(G1)        D9:" IN
B10:"#TEETH 1
B11:WT (G2)       D11:" OZ
B12:RAD (G2)      D12:" IN
B13:"-TEETH 2
B14:INERT(M)      D14:" OZ-SQ IN
B15:DEL RPM
B16:DEL TIME      D16:" SEC
A18:/ - *
/R:B18.D18
B20:INER(L)
   C20:.5*C6*((C7*C13/C10)^2)
   D20:" OZ SQ IN
B22:INER(G1)
   C22:.5*C8*((C9*C13/C10)^2)
   D22:" OZ SQ IN
B24:INER(G2)
   C24:.5*C11*C12*C12
   D24:" OZ SQ IN
B26: TORQUE
   C26:2*@PI*(C14+C20+C22+C24)*C15/(386*60*C16)
   D26:" OZ IN
```

Figure 7.14: *Spreadsheet Instructions: Motor Torque—Gear Drive*

LEAD COMPENSATION

PURPOSE

This spreadsheet allows us to calculate the design data for a lead compensation network given the necessary design parameters. The design equations are as follows:

$$\text{Transfer Function} = T_{\text{LEAD}}(s) = \frac{s + a}{s + b}$$

$$\frac{a}{b} = \left[\tan\left(\frac{90 - \theta}{2}\right)\right]^2$$

$$a = \omega_m \sqrt{\frac{a}{b}}$$

$$b = \frac{\omega_m}{\sqrt{\frac{a}{b}}}$$

$$\text{Attenuation @ } \omega_m = \sqrt{\frac{a}{b}}$$

$$\text{Atenuation @ } \omega_m \text{ (dB)} = 20 \log_{10} \sqrt{\frac{a}{b}}$$

where:

θ = Maximum phase lead, and
ω_m = frequency of maximum phase lead.

DATA ENTRY

We enter the frequency (in hertz) at which we wish to place the maximum phase lead. Next we enter the desired value of maximum phase lead. The program calculates the radian frequency and an initial ratio *a*/*b*. We can then use this value to specify the actual *a*/*b* ratio to be used in the design calculations. The design table is then produced, showing the values of *a* and *b* as well as the attenuation value and the attenuation in dB at the frequency of maximum phase lead.

EXAMPLE

We want to have the maximum phase lead at 0.54 Hz, and the desired amount of lead is 35 degrees. The design table calculates a radian frequency of 3.39 rad/sec and an *a/b* ratio of 0.27. We then decide to use an *a/b* value of 0.30 and enter this value into the spreadsheet. The resulting design data is presented in the table shown in Figure 7.15. Instructions for the spreadsheet are displayed in Figure 7.16.

```
        A            B            C            D
 1             LEAD COMENSATION
 2
 3             ENTER SPECIFICATIONS:
 4
 5             FREQ.(M)=       .54 HZ.
 6
 7             PHASE(M)=        35 DEGREES
 8
 9               (A/B)`=        .3
10
11  ***************************************
12
13             FREQ(M)  = 3.392920RAD./SEC
14               (A/B)  = .2709901
15
16                    A= 1.858379
17                    B= 6.194596
18             ATTEN@FM= .5477226
19             ATTEN@FM= -5.22879 D.B.
```

Figure 7.15: *Lead Compensation Spreadsheet*

```
B1:LEAD COMP      C1:ENSATION
B3:ENTER SPE      C3:CIFICATIO      D3:NS:
B5:FREQ.(M) =     D5:" HZ.
B7:PHASE(M) =     D7:" DEGREES
B9:"  (A/B)' =
A11:/- *
/R:B11.D11
B13:FREQ(M)  =
   C13:2*@PI*C5 D13:RAD./SEC
```

Figure 7.16: *Spreadsheet Instructions: Lead Compensation*

```
B14:"   (A/B) =
   C14:(@TAN(((90-C7)*@PI)/(2*180)))^2
B16:"      A=
   C16:@SQRT(C9)*C13
B17:"      B=
   C17:+C13/(@SQRT(C9))
B18:"ATTEN@FM=
   C18:@SQRT(C9)
B19:"ATTEN@FM=
   C19:20*@LOG10(C18)                D19:" D.B.
```

Figure 7.16, Continued

8

Physics

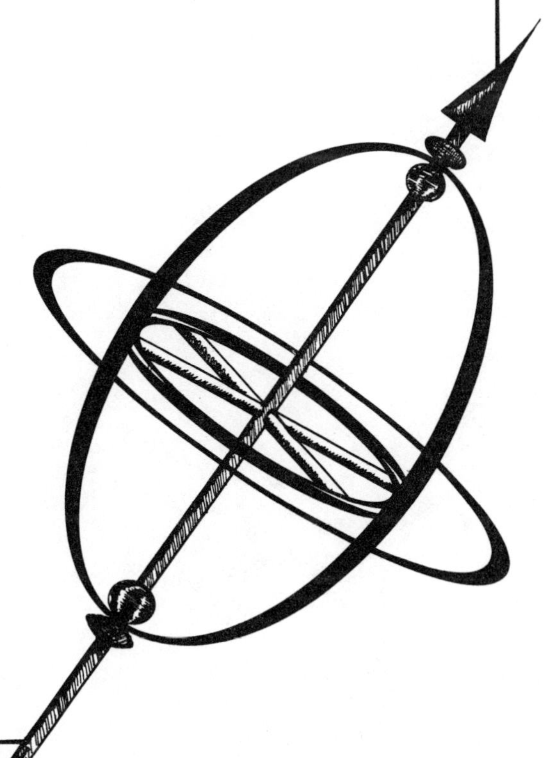

EARTH SATELLITE ORBIT

PURPOSE

In this application, we develop a spreadsheet to calculate Earth satellite orbital parameters, given a specified orbital period and an increment in that period. A second spreadsheet is also provided, to calculate orbital parameters given an orbital height above the Earth's surface and an increment in that height. The following equations are used:

$$\text{Height} = \sqrt[3]{2494 \times T^2} - 3964$$

$$\text{Velocity} = \frac{2\pi(\text{Height} + 3964)}{T}$$

$$\text{Period} = T = \sqrt{0.000401(\text{Height} + 3964)}$$

DATA ENTRY

If you know the orbital period, enter that value as well as a chosen value for the increment in period to obtain the orbital height and velocity as a function of period. If you know the orbital height, enter that value as well as a chosen value for the increment in height to obtain the orbital period and velocity as a function of height. The second spreadsheet appears in columns F through H.

EXAMPLE

We wish to determine the height and velocity of an earth satellite having a synchronous orbit (one day = 86,000 sec). We enter 80,000 sec as the initial period, and 1000 sec as the incremental step. Ten steps are calculated, and the height corresponding to 86,000 sec is found to be 22,458 miles. We also want to find the period and velocity of an orbit having an altitude ranging from 1 to 10 miles in 1-mile steps. The results are shown in the two tables of Figure 8.1. Velocity values in the right-hand column appear to be constant because they are rounded to two places. The spreadsheet instructions are shown in Figure 8.2.

	A	B	C	D	E	F	G	H
1		SATELLITE	ORBIT					
2								
3								
4		ENTER PAR	AMETERS:					
5								
6		PERIOD	80000.00	SEC		HEIGHT	1.00	MILE
7								
8		DEL PER	1000.00	SEC		DEL HGT	1.00	MILE
9								
10	**********	**********	**********	**********	**********	**********	**********	**********
11								
12		PERIOD	HEIGHT	VELOCITY		HEIGHT	PERIOD	VELOCITY
13		SEC	MILE	MI/SEC		MILE	SEC	MI/SEC
14								
15		80000.00	21214.25	1.98		1.00	4999.62	4.98
16		81000.00	21423.63	1.97		2.00	5001.51	4.98
17		82000.00	21632.15	1.96		3.00	5003.40	4.98
18		83000.00	21839.83	1.95		4.00	5005.30	4.98
19		84000.00	22046.68	1.95		5.00	5007.19	4.98
20		85000.00	22252.70	1.94		6.00	5009.08	4.98
21		86000.00	22457.92	1.93		7.00	5010.97	4.98
22		87000.00	22662.35	1.92		8.00	5012.87	4.98
23		88000.00	22866.00	1.92		9.00	5014.76	4.98
24		89000.00	23068.87	1.91		10.00	5016.65	4.98

Figure 8.1: *Satellite Orbit Spreadsheet*

```
/GF$
B1:SATELLITE        C1:" ORBIT
B4:ENTER PAR        C4:AMETERS:
B6:PERIOD           D6:" SEC
F6:HEIGHT           H6:" MILE
B8:DEL PER          D8:" SEC
F8:DEL HGT          H8:" MILE
A10:/-*
/R:B10.H10
B12:/FR PERIOD      B13:/FR SEC
```

Figure 8.2: *Spreadsheet Instructions: Satellite Orbit*

```
   B15: +C6
   B16: +B15+C8
   /R:B17.B24:RN
C12:/FR HEIGHT      C13:/FR MILE
   C15:2494*B15*B15^(1/3)-3964
   /R:C16.C24:RR
D12:/FR VELOCITY    D13:/FR MI/SEC
   D15:2*@PI*(C15+3964)/B15
   /R:D16.D24:RR
F12:/FR HEIGHT      F13:/FR MILE
   F15: +G6
   F16: +F15+G8
   /R:F17.F24:RN
G12:/FR PERIOD      G13:/FR SEC
   G15:.000401*((3964+F15)^3)^(1/2)
   /R:G16.G24:R
H12:/FR VELOCITY    H13:/FR MI/SEC
   H15:2*@PI*(F15+3964)/G15
   /R:H16.H24:RR
```

Figure 8.2, Continued

PROJECTILE MOTION

PURPOSE

We use this spreadsheet to calculate the spatial position of a projectile in two-dimensional space as a function of time. The initial conditions of the projectile are its launch velocity and angle. The equations of motion are:

$$x = v_x t$$

$$y = v_y t - \frac{1}{2} g t^2$$

where

$$v_x = v_0 \sin \alpha$$

$$v_y = v_0 \cos \alpha$$

and

$$g = 9.8 \text{ M/sec}^2$$

Metric units are used in this spreadsheet.

DATA ENTRY

We enter the initial velocity, initial angle, observation time, and the increment in time as information input to the spreadsheet. The initial x and y velocity components are calculated, as well as ten values in time of the x and y positions in space.

EXAMPLE

We want to determine the trajectory of a ball having an initial velocity of 25 m/sec and an angle of departure of 45 degrees. We choose to observe the event at zero time and in steps of 0.5 second. The results are displayed in the table of Figure 8.3. The spreadsheet instructions are shown in Figure 8.4.

```
          A          B          C          D
 1              PROJECTILE MOTION
 2
 3              ENTER PARAMETERS:
 4
 5              INIT VEL      25.00 M/SEC
 6
 7              INIT ANG      45.00 DEG
 8
 9              OBS TIME       0.00 SEC
10
11              DEL TIME       0.50 SEC
12
13  ************************************
14
15              VX= 17.68         VY=      17.68
16
17                   TIME     POS X      POS Y
18
19                   0.00      0.00       0.00
20                   0.50      8.84       7.61
21                   1.00     17.68      12.78
22                   1.50     26.52      15.49
23                   2.00     35.36      15.76
24                   2.50     44.19      13.57
25                   3.00     53.03       8.93
26                   3.50     61.87       1.85
27                   4.00     70.71      -7.69
28                   4.50     79.55     -19.68
```

Figure 8.3: *Projectile Motion Spreadsheet*

```
/GF$
B1:PROJECTIL      C1:E MOTION
B3:ENTER PAR      C3:AMETERS:
B5:INIT VEL       D5:" M/SEC
B7:INIT ANG       D7:" DEG
B9:OBS TIME       D9:" SEC
B11:DEL TIME      D11:" SEC
A13:/-*
/R:B13.D13
A15:/FR VX=
   B15: +C5*@COS(C7*@PI/180)
C15:/FR VY=
```

Figure 8.4: *Spreadsheet Instructions: Projectile Motion*

```
    D15: +C5*@SIN(C7*@PI/180)
B17:/FR TIME
    B19: +C9
    B20: +B19+C11
    /R:B21.B28:RN
C17:/FR POS X
    C19: +B19*B15
    /R:C20.C28:RN
D17:/FR POS Y
    D19: +D15*B19-(4.9*(B19^2))
    /R:D20.D28:NRR
```

Figure 8.4, Continued

PHYSICAL PENDULUM

PURPOSE

This spreadsheet allows you to determine the period of oscillation of a physical pendulum, given the distance between its pivot point and its center of gravity, its weight, and its moment of inertia. It also allows you to determine the distance given the period. An incremental variation from the nominally specified period or distance allows us to examine a range of results. The equations used are shown below:

$$\text{Period} = T = 2\pi\sqrt{\frac{I}{WD}}$$

$$\text{Distance} = D = \frac{I}{W\left(\frac{T}{2\pi}\right)^2}$$

where

I = Moment of Inertia, and
W = Weight.

DATA ENTRY

To determine the period of a pendulum, enter the weight and moment of inertia as well as an initial distance and an increment for distance. The table produces a range of values for distance and a corresponding range of period values. To determine the distance between the pivot point and the center of gravity, enter the weight and moment of inertia as well as an initial period and an increment for period. The table produces a range of period values and a corresponding range of distance values.

EXAMPLE

You wish to find the period of oscillation of a physical pendulum weighing 3.22 lb and having a moment of inertia of 0.807 slug-ft^2 for a distance that ranges from 2.2 ft upward in 0.05 ft increments. Also, you want to know the distance for a period that ranges from 1.8 sec upward in .05 sec increments. The design table displays the results for both sets of calculations, shown in Figure 8.5. The spreadsheet instructions are presented in Figure 8.6.

```
          A          B            C          D
 1                 PHYSICAL PENDULUM
 2
 3                 ENTER PARAMETERS:
 4
 5                 WEIGHT           3.22 LBS
 6                 MO INERT         0.81 SG-SQ FT
 7                 PERIOD           1.80 SEC
 8                 DEL PER          0.05 SEC
 9                 DISTANCE         2.20 FT
10                 DEL DIST         0.05 FT
11
12  **********************************************
13
14   DISTANCE     PERIOD      PERIOD  DISTANCE
15
16       2.20       2.12        1.80      3.05
17       2.25       2.10        1.85      2.89
18       2.30       2.07        1.90      2.74
19       2.35       2.05        1.95      2.60
20       2.40       2.03        2.00      2.47
21       2.45       2.01        2.05      2.35
22       2.50       1.99        2.10      2.24
23       2.55       1.97        2.15      2.14
24       2.60       1.95        2.20      2.04
25       2.65       1.93        2.25      1.95
26       2.70       1.91        2.30      1.87
```

Figure 8.5: *Physical Pendulum Spreadsheet*

```
/GF$
B1:PHYSICAL         C1:PENDULUM
B3:ENTER PAR        C3:AMETERS:
B5:WEIGHT           D5:" LBS
B6:MO INERT         D6:" SG-SQ FT
B7:PERIOD           D7:" SEC
B8:DEL PER          D8:" SEC
B9:DISTANCE         D9:" FT
B10:DEL DIST        D10:" FT
A12:/-*
/R:B12.D12
A14:/FR DISTANCE
   A16:+C9
   A17:+A16+C10
   /R:A18.A26:RN
B14:/FR PERIOD
   B16:2*@PI*@SQRT(C6/(C5*A16))
   /R:B17.B26:NNR
C14:/FR PERIOD
   C16:+C7
   C17:+C16+C10
   /R:C18.C26:RN
D14:/FR DISTANCE
   D16:+C6/(C5*((C16/(2*@PI))^2))
   /R:D17.D26:NNR
```

Figure 8.6: *Spreadsheet Instructions: Physical Pendulum*

SIMPLE PENDULUM

PURPOSE

This spreadsheet allows you to determine the oscillation period of a simple pendulum given its length, and its length given the period of oscillation. A variation from the nominal specified length or period provides a range of results for examination. The equations used are shown below:

$$\text{Period} = 2\pi\sqrt{\text{Length}/g}$$

and

$$\text{Length} = g\left(\frac{\text{Period}}{2\pi}\right)^2$$

where

$$g = 32.2 \text{ ft/sec}^2$$

DATA ENTRY

To determine the period of a pendulum, enter an initial length and a length increment. The table produces twelve period values, corresponding to twelve different lengths. To determine the length of a pendulum, enter an initial period and a period increment. The table then produces twelve length values, corresponding to twelve different periods.

EXAMPLE

You want to know the oscillation period of a pendulum whose length may range from 10 to 20 feet. A value of 10 ft is entered for the length, and 1 ft is the length increment. Also, you want to find the length of a pendulum whose period of oscillation may range from 3.5 sec to 4.5 sec in incremental steps of 0.1 sec. The design table generates the results shown in Figure 8.7. The first set of values appears in columns A and B, the second in columns C and D. The instructions for the spreadsheet are presented in Figure 8.8.

```
        A          B          C          D
 1              SIMPLE    PENDULUM
 2
 3              ENTER PARAMETERS:
 4
 5              LENGTH         10.00 FT
 6              DEL LEN         1.00 FT
 7              PERIOD          3.50 SEC
 8              DEL PER         0.10 SEC
 9
10  ************************************
11
12     LENGTH     PERIOD     PERIOD     LENGTH
13
14      10.00       3.50       3.50       9.99
15      11.00       3.67       3.60      10.57
16      12.00       3.84       3.70      11.17
17      13.00       3.99       3.80      11.78
18      14.00       4.14       3.90      12.40
19      15.00       4.29       4.00      13.10
20      16.00       4.43       4.10      13.71
21      17.00       4.57       4.20      14.39
22      18.00       4.70       4.30      15.08
23      19.00       4.83       4.40      15.79
24      20.00       4.95       4.50      16.52
25      21.00       5.07       4.60      17.26
```

Figure 8.7: *Simple Pendulum Spreadsheet*

```
/GF$
B1:SIMPLE          C1:PENDULUM
B3:ENTER PAR       C3:AMETERS:
B5:LENGTH          D5:" FT
B6:DEL LEN         D6:" FT
B7:PERIOD          D7:" SEC
B8:DEL PER         D8:" SEC
A10:/-*
/R:B10.D10
A12:/FR LENGTH
   A14:+C5
   A15:+A14+C6
   /R:A16.A25:RN
B12:/FR PERIOD
```

Figure 8.8: *Spreadsheet Instructions: Simple Pendulum*

```
   B14:2*@PI*@SQRT(A14/32.2)
   /R:B15.B25:R
C12:/FR PERIOD
   C14:+C7
   C15:+C14+C8
   /R:C16.C25:RN
D12:/FR LENGTH
   D14:32.2*((C14/(2*@PI))^2)
   /R:D15.D25:R
```

Figure 8.8, Continued

THIN LENS EQUATION

PURPOSE

Given the focal length of a thin lens and an object height, this spreadsheet computes the image height, magnification, and image distance as a function of the object distance. The equations used are:

$$\text{Image distance} = S_i = \frac{S_o F}{S_o - F}$$

$$\text{Magnification} = M = -\frac{S_i}{S_o}$$

$$\text{Image height} = Y_i = MY_o$$

where:

S_o = Object distance,
Y_o = Object height, and
F = Focal length.

DATA ENTRY

You first enter the focal length and object height, then an initial object distance and an object distance step size. The resulting table displays the image distance, magnification, and image height as a function of ten different values of object distance. Note that errors in calculation (such as division by zero) can be produced as a result of using certain combinations of optical values in the spreadsheet.

EXAMPLE

You have a thin lens of focal length 25 cm and want to examine the optical parameters of a 3 cm object at various distances from the lens. You start at 50 cm and increment in 5 cm steps. The results are generated and displayed in the table of Figure 8.9. The spreadsheet instructions are given in Figure 8.10.

	A	B	C	D
1		THIN LENS EQUATION		
2				
3		ENTER PARAMETERS:		
4				
5		FOCAL LT=	25.00	
6		OBJ. HGT=	3.00	
7		OBJ. DST=	50.00	
8		DEL OD =	5.00	
9				
10	**********	**********	**********	**********
11				
12	OBJ.DIS	IMAGE DIS	MAG	IMAGE HT
13				
14	50.00	50.00	-1.00	-3.00
15	55.00	45.83	-0.83	-2.50
16	60.00	42.86	-0.71	-2.14
17	65.00	40.63	-0.63	-1.88
18	70.00	38.89	-0.56	-1.67
19	75.00	37.50	-0.50	-1.50
20	80.00	36.36	-0.45	-1.36
21	85.00	35.42	-0.42	-1.25
22	90.00	34.62	-0.38	-1.15
23	95.00	33.93	-0.36	-1.07

Figure 8.9: *Thin Lens Spreadsheet*

```
/GF$
B1:THIN LENS       C1:" EQUATION
B3:ENTER PAR       C3:AMETERS:
B5:FOCAL LT=
B6:OBJ. HGT=
B7:OBJ. DST=
B8:DEL OD  =
A10:/-*
/R:B10.D10
A12:OBJ.DIS
   A14:+C7
   A15:+A14+C8
   /R:A16.A23:RN
B12:IMAGE DIS
   B14:+A14*C5/(A14-C5)
   /R:B15.B23:RNRN
C12:"  MAG
   C14:-B14/A14
   /R:C15.C23:RR
D12:" IMAGE HT
   D14:+C14*C6
   /R:D15.D23:RN
```

Figure 8.10: *Spreadsheet Instructions: Thin Lens*

THICK LENS FOCAL LENGTH

PURPOSE

Given the two radii of a thick lens, along with the index of refraction of the material and a range of lens thicknesses, you can use this spreadsheet to determine the focal length and the locations of the first and second principal planes. The equations used are listed below:

$$\text{Focal length} = F = \frac{1}{(n-1)\left[\frac{1}{R_1} - \frac{1}{R_2} + \frac{(n-1)d}{nR_1R_2}\right]}$$

$$\text{First principal plane location} = -\frac{F(n-1)d}{nR_2}$$

$$\text{Second principal plane location} = -\frac{F(n-1)d}{nR_1}$$

where

d = lens thickness, and
n = index of refraction.

DATA ENTRY

You first enter the two radii, then the index of refraction, and finally the lens thickness along with the thickness increment. The resulting table shows the focal length and the first and second principal planes as a function of ten different lens thicknesses. You should note that this spreadsheet is calculated across by rows rather than down by columns.

EXAMPLE

You wish to examine the effects of variations in lens thickness for a design having a radius 1 of 60 mm and a radius 2 of −40 mm. The index of refraction for the glass is 1.5, the initial thickness is 10 mm, and the thickness increment is 2 mm. The resulting table is displayed in Figure 8.11. Instructions for this spreadsheet are given in Figure 8.12.

	A	B	C	D
1		THICK LENS FOCAL		
2		LENGTH		
3				
4		ENTER PARAMETERS:		
5				
6		RADIUS 1=	60.00	
7		RADIUS 2=	-40.00	
8		INDEX =	1.50	
9		THICK =	10.00	
10		DEL.THCK=	2.00	
11				
12	**********	**********	**********	**********
13				
14	THICKNESS	FOC.LGT.	PLANE 1	PLANE 2
15				
16	10.00	49.66	4.14	-2.76
17	12.00	50.00	5.00	-3.33
18	14.00	50.35	5.87	-3.92
19	16.00	50.70	6.76	-4.51
20	18.00	51.06	7.66	-5.11
21	20.00	51.43	8.57	-5.71
22	22.00	51.80	9.50	-6.33
23	24.00	52.17	10.43	-6.96
24	26.00	52.55	11.39	-7.59
25	28.00	52.94	12.35	-8.24

Figure 8.11: *Thick Lens Spreadsheet*

```
/GOR
/GF$
B1:THICK LEN        C1:S FOCAL
B2:LENGTH
B4:ENTER PAR        C4:AMETERS:
B6:RADIUS 1 =
B7:RADIUS 2 =
```

Figure 8.12: *Spreadsheet Instructions: Thick Lens*

```
B8:INDEX  =
B9:THICK  =
B10:DEL.THCK=
A12:/ - *
/R:B12.D12
A14:THICKNESS
   A16: +C9
   A17: +A16+C10
   /R:A18.A25:RN
B14:" FOC.LGT.
   B16:1/((C8-1)*((1/C6)-(1/C7)+((C8-1)*A16/(C8*C6*C7)))
   /R:B17.B25:NNNNRNNN
C14:"  PLANE 1
   C16: -B16*(C8-1)*A16/(C7*C8)
   /R:C17.C25:RNRNN
D14:"  PLANE 2
   D16: -B16*(C8-1)*A16/(C6*C8)
   /R:D17.D25:RNRNN
```

Figure 8.12, Continued

COMPOUND THIN LENS

PURPOSE

We use this spreadsheet to calculate the image distance and magnification for a compound set of two thin lenses. The lens system parameters are the two focal lengths and the distance between the lenses. We vary the object distance from an initial value through a range of ten values. Values for the front and back focal lengths are also included as results. The equations used are:

$$\text{Image distance} = S_i = \frac{F_2 d - [F_1 F_2 S_o/(S_o - F_1)]}{d - F_2 - [F_1 S_o/(S_o - F_1)]}$$

$$\text{Magnification} = M = \frac{F_1 S_i}{d(S_o - F_1) - S_o F_1}$$

$$\text{Front focal length} = \frac{F_1(d - F_2)}{d - (F_1 + F_2)}$$

$$\text{Back focal length} = \frac{F_2(d - F_1)}{d - (F_1 + F_2)}$$

where:

d = distance between lenses,
S_o = object distance,

and

F_1, F_2 are focal lengths.

DATA ENTRY

First we enter the two focal lengths, then the distance between the two lenses, and finally the object distance, along with the object distance increment. The spreadsheet table shows the image distance and magnification as functions of ten different object distances. The front focal length and back focal length values are also calculated and displayed. Note that errors in calculation (such as division by zero) can be produced as a result of using certain combinations of optical values in the spreadsheet.

EXAMPLE

We wish to determine the effects of variations in object distance for a lens system design having a focal length 1 of 2 cm and a focal length 2 of 5 cm. The distance between the lenses is 10 cm, the initial object distance is 2.8 cm, and the distance increment is 0.05 cm. The resulting spreadsheet is shown in Figure 8.13. Instructions for the spreadsheet are given in Figure 8.14.

```
          A          B          C          D
 1               COMPOUND THIN LENS
 2
 3               ENTER PARAMETERS:
 4
 5               FOC LT 1=       2.00
 6               FOC LT 2=       5.00
 7               DISTANCE=      10.00
 8               OBJ DIST=       2.80
 9               DEL OD  =       0.05
10
11  ****************************************
12
13  FFL =             3.33   BFL =     13.33
14
15                 OBJ DST   IMG DST     MAG
16
17                    2.80     -7.50   -6.25
18                    2.85     -9.66   -6.90
19                    2.90    -12.31   -7.69
20                    2.95    -15.65   -8.70
21                    3.00    -20.00  -10.00
22                    3.05    -25.88  -11.76
23                    3.10    -34.29  -14.29
24                    3.15    -47.27  -18.18
25                    3.20    -70.00  -25.00
26                    3.25   -120.00  -40.00
```

Figure 8.13: *Compound Thin Lens Spreadsheet*

```
/GF$
B1:COMPOUND      C1:THIN LENS
B3:ENTER PAR     C3:AMETERS:
B5:FOC LT 1=
```

Figure 8.14: *Spreadsheet Instructions: Compound Thin Lens*

```
B6:FOC LT 2=
B7:DISTANCE=
B8:OBJ DIST=
B9:DEL OD  =
A11:/- *
/R:B11.D11
A13:FFL =
   B13: +C5*(C7-C6)/(C7-(C5+C6))
C13:"  BFL =
   D13: +C6*(C7-C5)/(C7-(C5+C6))
B15:/FR OBJ DST
   B17: +C8
   B18: +B17+C9
   /R:B19.B26:RN
C15:/FR IMG DST
   C17:(C6*C7-(C5*C6*B17/(B17-C5)))/(C7-C6-(C5*B17/(B17-C5)))
   /R:C18.C26:NNN NRR NNN NRRN
D15:/FR MAG
   D17: +C5*C17/((C7*(B17-C5))-(B17*C5))
   /R:D18.D26:NRNRNRN
```

Figure 8.14, Continued

9 Mechanical and Civil Engineering

FLUID FLOW PRESSURE LOSS

PURPOSE

In this application we develop a spreadsheet to compute the loss in fluid pressure due to fluid flow from elevated tanks. Given the tank height, fluid density, and a range of flow velocities, the static fluid pressure is computed as a function of ten different velocity values. From Bernoulli's equation, the static pressure is given by:

$$\text{Pressure} = \rho\left(g\,h - \frac{V^2}{2}\right)$$

where:

ρ = fluid density,
h = height,
V = fluid output velocity, and
g = 32.2 ft/sec^2.

DATA ENTRY

First we enter the tank height, then the fluid density. Next we enter the initial velocity, along with the incremental step size in velocity. The spreadsheet table produces the static pressure as a function of ten different values of fluid velocity.

EXAMPLE

A water tank is 100 feet high and the density of water is known to be 1.94 slug/ft^3. We are interested in determining the range of static pressures for flow rates ranging from 0 to 45 ft/sec in steps of 5 ft/sec. The resulting spreadsheet is shown in Figure 9.1. The instructions for the spreadsheet are given in Figure 9.2.

```
          A           B           C           D

 1                 FLUID       FLOW
 2
 3
 4                 ENTER PARAMETERS:
 5
 6                 HEIGHT          100.00 FT
 7
 8                 DENSITY           1.94 SLUG/FT3
 9
10                 INIT VEL          0.00 FT/S
11
12                 DEL VEL           5.00 FT/S
13
14   ************************************
15
16                 VELOCITY      PRESS
17
18                     0.00      43.38
19                     5.00      43.21
20                    10.00      42.71
21                    15.00      41.86
22                    20.00      40.69
23                    25.00      39.17
24                    30.00      37.32
25                    35.00      35.13
26                    40.00      32.60
27                    45.00      29.74
```

Figure 9.1: *Fluid Flow Spreadsheet*

```
/GF$
B1:FLUID            C1:FLOW
B4:ENTER PAR        C4:AMETERS:
B6:HEIGHT           D6:" FT
B8:DENSITY          D8:" SLUG/FT3
B10:INIT VEL        D10:" FT/S
B12:DEL VEL         D12:" FT/S
A14:/-*
/R:B14.D14
B16:/FR VELOCITY
   B18: +C10
   B19: +B18+C12
```

Figure 9.2: *Spreadsheet Instructions: Fluid Flow*

```
   /R:B20.B27:RN
C16:/FR PRESS
   C18: +C8*(32.2*C6-(.5*B18*B18))/144
   /R:C19.C27:NNRR
```

Figure 9.2, Continued

VENTURI FLOW METER

PURPOSE

We wish to determine the rate of fluid flow by measuring the difference in height of liquid between two manometers. This spreadsheet allows us to determine the flow rate as a function of height difference, given two cross-sectional areas. The venturi meter equation is:

$$\text{Flow rate} = R = A_1 \sqrt{\frac{2gh}{(A_1/A_2)^2 - 1}}$$

where:

A_1 is the (larger) pipe cross-sectional area,
A_2 is the (smaller) meter cross-sectional area,
h = height difference,
g = 32.2 ft/sec^2, and
R is measured in cu ft/sec.

DATA ENTRY

We enter the larger area as AREA 1 and the smaller area as AREA 2. We then define an initial height difference as well as an incremental change in height. The spreadsheet table produces the flow rate as a function of ten different height values.

EXAMPLE

We are interested in a flow meter design having a pipe area of 0.6 ft^2 and a flow meter area of 0.2 ft^2. We wish to determine the relationship between height values and flow rates for this design. Our initial value of height difference is 0.2 ft, and the step size is 0.01 ft. These values are entered into the spreadsheet, with the results shown in the table of Figure 9.3. Instructions for the spreadsheet are given in Figure 9.4.

	A	B	C	D
1		VENTURI FLOW METER		
2				
3		ENTER PARAMETERS:		
4				
5		AREA 1	0.60	SQ FT
6				
7		AREA 2	0.20	SQ FT
8				
9		HEIGHT	0.20	FT
10				
11		DEL HT	0.01	FT
12				
13	************************************			
14				
15		HEIGHT	FLO RATE	
16				
17		0.20	0.76	
18		0.21	0.78	
19		0.22	0.80	
20		0.23	0.82	
21		0.24	0.83	
22		0.25	0.85	
23		0.26	0.87	
24		0.27	0.88	
25		0.28	0.90	
26		0.29	0.92	

Figure 9.3: *Venturi Flow Meter Spreadsheet*

```
/GF$
B1:VENTURI F         C1:LOW METER
B3:ENTER PAR         C3:AMETERS:
B5:AREA 1            D5:" SQ FT
B7:AREA 2            D7:" SQ FT
B9:HEIGHT            D9:" FT
B11:DEL HT           D11:" FT
A13:/ - *
/R:B13.D13
B15:/FR HEIGHT
   B17: +C9
   B18: +B17+C11
   /R:B19.B26:RN
C15:/FR FLO RATE
   C17: +C5*@SQRT(2*32.2*B17/(((C5/C7)^2)-1))
   /R:C18.C26:NRNN
```

Figure 9.4: *Spreadsheet Instructions: Venturi Flow Meter*

MECHANICAL VIBRATIONS

PURPOSE

This spreadsheet allows you to determine the critical damping coefficient, the undamped natural frequency, the damping ratio, and the damped frequency of oscillation for a mechanical system subject to vibration, given the following information: the weight of the object, the spring constant, and the initial damping coefficient. A step change in that coefficient produces a series of values for the damping ratio and damped frequency. The design equations are shown below:

$$\text{Critical Damping Coefficient} = CD_{CO} = 2\sqrt{k \times W/g}$$

$$\text{Natural Frequency} = F_n = \frac{1}{2\pi}\sqrt{\frac{kg}{W}}$$

$$\text{Damping Ratio} = DR = \frac{D_{CO}}{CD_{CO}}$$

$$\text{Damped Frequency} = F_n\sqrt{1 - DR^2}$$

where:

W = object weight,
k = spring constant,
D_{CO} = damping coefficient, and
g = 386.4 in./sec^2.

DATA ENTRY

You enter the object weight, the spring constant, an initial damping coefficient value, and an incremental step size for the damping coefficient values. The spreadsheet produces the critical damping coefficient, natural undamped frequency, and the damping ratio and damped frequency as a function of ten different values for the damping coefficient.

EXAMPLE

You want to examine the effects of various damping coefficients, ranging from 0.1 to 1.0 in steps of 0.1, on a mechanical system having a weight of 12 pounds and a spring constant of 20 pounds/inch. These values are entered into the spreadsheet, with the results shown in the table of Figure 9.5. Instructions for the spreadsheet are given in Figure 9.6.

```
         A          B          C          D
 1            MECHANICAL VIBRATIONS
 2
 3            ENTER PARAMETERS:
 4
 5            WEIGHT         12.00 LBS
 6
 7            SPRG CONS      20.00 LBS/IN
 8
 9            DAMP CO         0.10
10
11            DEL DCO         0.10
12
13  ***************************************
14
15          CRIT DCO       1.58
16
17          NAT FREQ       4.04 HZ
18
19              DAMP CO DAMP RAT DAMP FRE
20
21                 0.10      0.06      4.03
22                 0.20      0.13      4.01
23                 0.30      0.19      3.97
24                 0.40      0.25      3.91
25                 0.50      0.32      3.83
26                 0.60      0.38      3.74
27                 0.70      0.44      3.62
28                 0.80      0.51      3.48
29                 0.90      0.57      3.32
30                 1.00      0.63      3.12
```

Figure 9.5: *Mechanical Vibrations Spreadsheet*

```
/GF$
B1:MECHANICA     C1:L VIBRATI     D1:ONS
B3:ENTER PAR     C3:AMETERS:
B5:WEIGHT        D5:" LBS
B7:SPRG CONS     D7:" LBS/IN
B9:DAMP CO
B11:DEL DCO
A13:/ – *
/R:B13.D13
B15:CRIT DCO
```

Figure 9.6: *Spreadsheet Instructions: Mechanical Vibrations*

```
    C15:2*@SQRT(C7*C5/386.4)
  B17:NAT FREQ
    C17:.1592*@SQRT(C7*386.4/C5)
    D17:" HZ
  B19:/FR DAMP CO
    B21:+C9
    B22:+B21+C11
    /R:B23.B30:RN
  C19:/FR DAMP RAT
    C21:+B21/C15
    /R:C22.C30:RN
  D19:/FR DAMP FRE
    D21:+C17*@SQRT(1-((C21)^2))
    /R:D22.D30:NR
```

Figure 9.6, Continued

SPECIAL SECTIONS

PURPOSE

This application produces a table showing the area, x-inertia, and y-inertia for rectangular, triangular, circular, concentric circular, and elliptical cross-sections. The two parameters defining the planar dimensions a and b and the associated equations in each case are illustrated below:

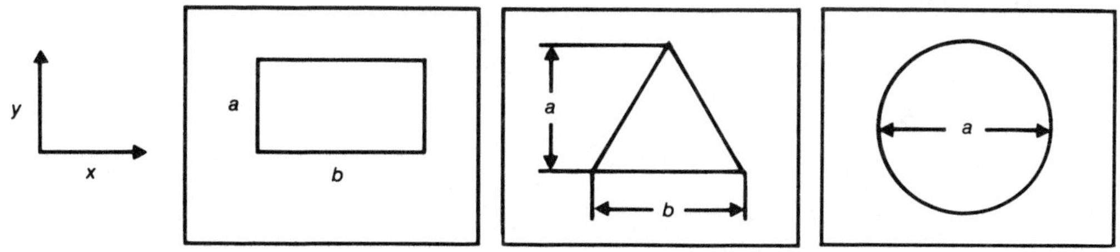

$A = ab$

$I_x = a^3b/12$

$I_y = ab^3/12$

$A = ab/2$

$I_x = a^3b/36$

$I_y = ab^3/36$

$A = \pi a^2/4$

$I_x = \pi a^4/64$

$I_y = \pi a^4/64$

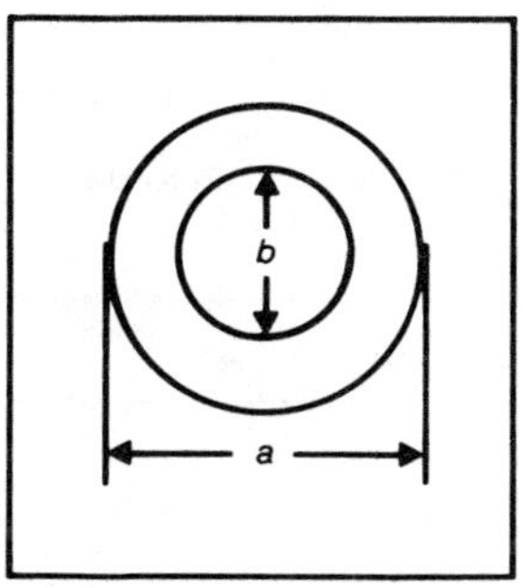

$A = \pi (a^2 - b^2)/4$

$I_x = \pi (a^4 - b^4)/64$

$I_y = \pi (a^4 - b^4)/64$

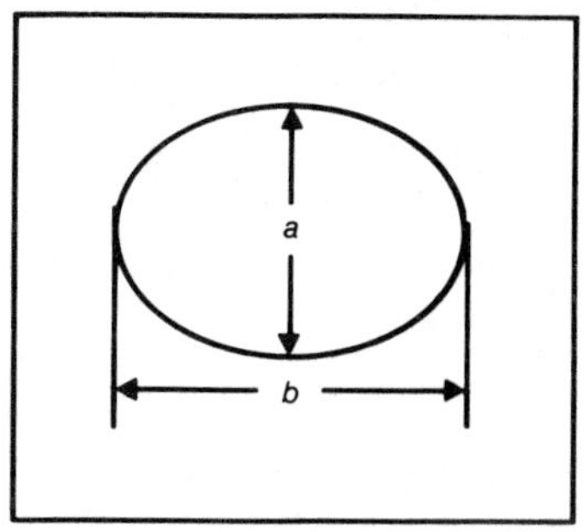

$A = \pi ab/4$

$I_x = \pi a^3b/64$

$I_y = \pi ab^3/64$

DATA ENTRY

You enter the dimensions of *a* and *b* (which are the same for all the sections), and the resulting table displays the data of interest for each type of section. Note that this spreadsheet uses columns E and F.

EXAMPLE

You want to examine the properties of various sections having the dimensions $a = 40$ and $b = 20$. The resulting spreadsheet is shown in Figure 9.7, and the instructions for the spreadsheet are given in Figure 9.8.

```
           A          B         C         D         E         F
 1
 2              SPECIAL   SECTIONS
 3
 4
 5              ENTER PARAMETERS:
 6
 7                               A=        40
 8
 9                               B=        20
10
11 *******************************************************************
12
13                    RECT       TRI    CIRCLE   CON CIR   ELLIPSE
14
15 AREA                800       400      1257       942       628
16
17 X INERTIA        106666     35556    125664    117810     62832
18
19 Y INERTIA         26667      8889    125664    117810     15708
```

Figure 9.7: *Special Sections Spreadsheet*

```
/GFI
B2:SPECIAL         C2:SECTIONS
B5:ENTER PAR       C5:AMETERS:
C7:/FR A=
C9:/FR B=
A11:/-*
```

Figure 9.8: *Spreadsheet Instructions: Special Sections*

```
/R:B11.F11
A15:AREA          A17:X INERTIA        A19:Y INERTIA
B13:/FR RECT
   B15: +D7*D9
   B17:(D7^3)*D9/12
   B19: +D7*(D9^3)/12
C13:/FR TRI
   C15: +D7*D9/2
   C17:(D7^3)*D9/36
   C19: +D7*(D9^3)/36
D13:/FR CIRCLE
   D15:@PI*(D7^2)/4
   D17:@PI*(D7^4)/64
   D19:@PI*(D7^4)/64
E13:/FR CON CIR
   E15:@PI*((D7*D7)-(D9*D9))/4
   E17:@PI*((D7^4)-(D9^4))/64
   E19:@PI*((D7^4)-(D9^4))/64
F13:/FR ELLIPSE
   F15:@PI*D7*D9/4
   F17:@PI*(D7^3)*D9/64
   F19:@PI*D7*(D9^3)/64
```

Figure 9.8, Continued

FIN EFFICIENCY

PURPOSE

You can use this spreadsheet to compute the temperature difference, the fin efficiency, and the total heat flux of a straight cooling fin design, given the fin parameters. The basic design equations are shown below:

$$\text{Efficiency} = E = \frac{e^{+x} - e^{-x}}{e^{+x} + e^{-x}} \times \frac{100}{x}$$

$$x = \left(l + \frac{t}{2}\right)^{3/2} \sqrt{2V/(\mathrm{D} \times \mathrm{t} \times \mathrm{l})}$$

$$\text{Heat flux} = V\left\{(1 - Nt) + E \times N(2l + t)\right\}\Delta T$$

$$\Delta T = |T_o - T_f|$$

where:

- N = number of fins,
- D = Conduction coefficient,
- V = Convection coefficient,
- T_o = Fin temperature,
- T_f = Fluid temperature,
- t = Fin thickness, and
- l = Fin length.

The following diagram illustrates the dimensional information.

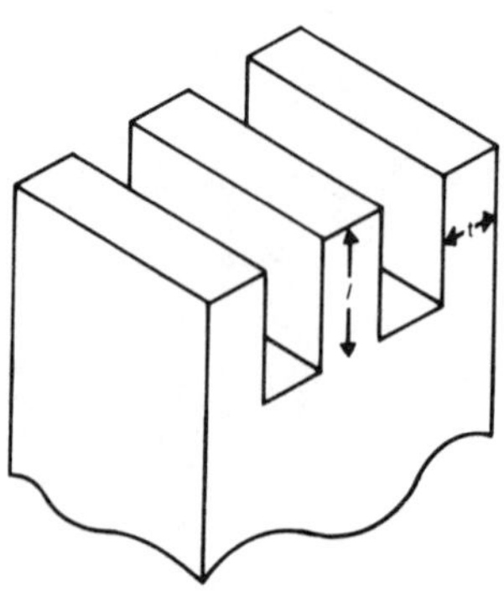

DATA ENTRY

You first enter the number of fins per unit of length, then the fin conduction coefficient and the fluid convection coefficient, the fin temperature and the fluid temperature, and finally the fin thickness, the fin length and the increment in fin length. The design table computes the temperature difference, and the efficiency and heat flux as functions of ten different fin lengths. Note that the column labeled with asterisks is used only for intermediate calculations of *x*. English units are used in the calculations.

EXAMPLE

You wish to examine the effects of various fin lengths on fin efficiency and heat flux. The number of fins is 10 per foot. The conduction coefficient for aluminum is 133 Btu/hr × °F × ft, and the convection coefficient for air is 50 Btu/hr × °F × ft. The fin temperature is 200 °F and the high-

```
          A           B           C           D

 1               FIN EFFICIENCY
 2
 3               ENTER PARAMETERS:
 4
 5
 6               NO OF FIN      10.00
 7               COND COEF     133.00
 8               CONV COEF      50.00
 9               FIN TEMP      200.00
10               FLU TEMP      120.00
11               FIN THICK       0.10
12               INIT LEN        0.10
13               LEN INC         0.10
14
15   ************************************
16
17               TEMP DIFF      80.00
18
19        FIN LEN       ***     EFF % HEAT FLX
20
21          0.10       0.15     99.30  4659.68
22          0.20       0.22     98.40  5306.66
23          0.30       0.30     97.12  5932.76
24          0.40       0.38     95.50  6531.63
25          0.50       0.46     93.58  7098.11
26          0.60       0.54     91.42  7628.33
27          0.70       0.61     89.06  8119.76
28          0.80       0.69     86.55  8571.12
29          0.90       0.77     83.93  8982.22
30          1.00       0.85     81.25  9353.85
```

Figure 9.9: *Fin Efficiency Spreadsheet*

est air temperature is 120 °F. The thickness of the fin is 0.1 inch and the starting fin length is 0.1 inch, incrementing in 0.1 inch steps. The results are displayed in Figure 9.9 (as shown on previous page). Instructions for this spreadsheet are given in Figure 9.10.

```
/GF$
B1:FIN EFFIC        C1:IENCY
B3:ENTER PAR        C3:AMETERS:
B6:NO OF FIN
B7:COND COEF
B8:CONV COEF
B9:FIN TEMP
B10:FLU TEMP
B11:FIN THICK
B12:INIT LEN
B13:LEN INC
A15:/-*
/R:B15.D15
B17:TEMP DIFF
   C17:@ABS(C9-C10)
A19:/FR FIN LEN
   A21:+C12
   A22:+A21+C13
   /R:A23.A30:RN
B19:" ***
B21:(A21+(C11/2))/12^(3/2)*@SQRT(288*C8/(C7*C11*A21))
   /R:B22.B30:RNNNNR
C19:/FR EFF %
   C21:100*((@EXP(B21)-@EXP(-B21))/((@EXP(B21)+@EXP(-B21))*B21
   /R:C22.C30:RRRRR
D19:/FR HEAT FLX
   D21:+C8*((1-(C6*C11/12))+(.01*C21*C6*((A21/6)+(C11/12))))*C17
   /R:D22.D30:NNNRNRNN
```

Figure 9.10: *Spreadsheet Instructions: Fin Efficiency*

BOLT TORQUE

PURPOSE

You can use this spreadsheet to compute a specified bolt's thread, collar, and total torques and its maximum shear stress as functions of load. The input information that you supply is the bolt parameter data and the initial load and load increment data. The design equations are as follows:

$$\text{Thread Torque} = T_T = \frac{L\,D_T}{2}\left[\frac{\tan\alpha + F_T/\cos\theta}{1 - F_T\tan\alpha/\cos\theta}\right]$$

$$\text{Collar torque} = C_T = F_C L \frac{D_C}{2}$$

$$\text{Total torque} = T_T + C_T$$

$$\text{Max. Shear Stress} = \sqrt{\left(\frac{2L}{\pi D_R^2}\right)^2 + \left(\frac{16T_T}{\pi D_R^3}\right)^2}$$

where:

L = bolt load,
D = mean thread diameter,
α = helix angle,
θ = one-half the thread angle,
F_T = coefficient of thread friction,
F_C = coefficient of collar friction,
D_C = collar diameter, and
D_R = root diameter of thread.

DATA ENTRY

You first enter the mean thread and root thread diameter data. Then enter the helix angle and one-half the thread angle. Next you enter the coefficient of thread friction, the collar diameter, and the coefficient of collar friction. Finally enter the initial load and the load increment. The design table produces the torque and stress information as a function of twelve different values of load. Note that the stress data is presented in column E.

EXAMPLE

You wish to determine the total torque and stress on a bolt subjected to loads in the range of 9500 to 10,600 lb. The bolt has a mean thread diameter of 0.9233 in., a root thread diameter of .8466 in., a helix thread angle of 2.47 degrees, a one-half thread-angle value of 30 degrees, a coefficient of thread friction of 0.15, a collar diameter of 1.25 in., and a coefficient of collar friction of 0.15. Your initial load is 9500 lb, incrementing in steps of 100 lb. The results are displayed in Figure 9.11. Instructions for this spreadsheet are given in Figure 9.12.

	A	B	C	D	E
1		BOLT	TORQUE		
2					
3		ENTER PARAMETERS:			
4					
5		M TH DIA	0.92	IN	
6		R TH DIA	0.85	IN	
7		H TH DIA	2.47	DEG	
8		.5 TH ANG	30.00	DEG	
9		C TH FRI	0.15		
10		COL DIA	1.25	IN	
11		C CR FRI	0.15		
12		IN LOAD	9500.00	LBS	
13		STP LOAD	100.00	LBS	
14					
15	**********	**********	**********	**********	**********
16					
17	LOAD	THRD TOR	COL TOR	TOT TOR	M SR STR
18					
19	9500.00	952.53	890.63	1843.15	11509.55
20	9600.00	962.56	900.00	1862.56	11630.70
21	9700.00	972.58	909.38	1881.96	11751.85
22	9800.00	982.61	918.75	1901.36	11873.01
23	9900.00	992.64	928.13	1920.76	11994.16
24	10000.00	1002.66	937.50	1940.16	12115.31
25	10100.00	1012.69	946.88	1959.56	12236.47
26	10200.00	1022.72	956.25	1978.97	12357.62
27	10300.00	1032.74	965.63	1998.37	12478.77
28	10400.00	1042.77	975.00	2017.77	12599.93
29	10500.00	1052.80	984.38	2037.17	12721.08
30	10600.00	1062.82	993.75	2056.57	12842.23

Figure 9.11: *Bolt Torque Spreadsheet*

```
/GF$
B1:BOLT            C1:TORQUE
B3:ENTER PAR       C3:AMETERS:
B5:M TH DIA        D5:" IN
B6:R TH DIA        D6:" IN
B7:H TH ANG        D7:" DEG
B8:".5 TH ANG      D8:" DEG
B9:C TH FRI
B10:COL DIA        D10:" IN
B11:C CR FRI
B12:IN LOAD        D12:" LBS
B13:STP LOAD       D13:" LBS
A15:/ - *
/R:B15.E15
A17:/FR
/R:B17.E17
A17:LOAD
   A19: +C12
   A20: +A19+C13
   /R:A21.A30:RN
B17:THRD TOR
   B19:.5*A19*C5*((@TAN(@PI*C7/180)+(C9/@COS(@PI*C8/180))/
   (1-((C9*@TAN(@PI*C7/180))/@COS(@PI*C8/180))))
   /R:B20.B30:RNNNNNNN
C17:COL TOR
   C19:.5*A19*C11*C10
   /R:C20.C30:RNN
D17:TOT TOR
   D19: +B19+C19
   /R:D20.D30:RR
E17:M SR STR
   E19:@SQRT(((2*A19/(@PI*(C6^2)))^2)+(((16*B19)/
   (@PI*(C6^3)))^2))
   /R:E20.E30:RNRN
```

Figure 9.12: *Spreadsheet Instructions: Bolt Torque*

SIMPLY SUPPORTED BEAMS

PURPOSE

This spreadsheet allows you to calculate the moment, y-deflection, slope, and shear at any point x along a simply supported beam of uniform cross-section. Problems modeled by any combination of point load, distributed load, or applied moment can be analyzed. You supply the beam parameters, the loading data, and the x coordinate for the point of interest. The beam geometry is shown in the diagram below:

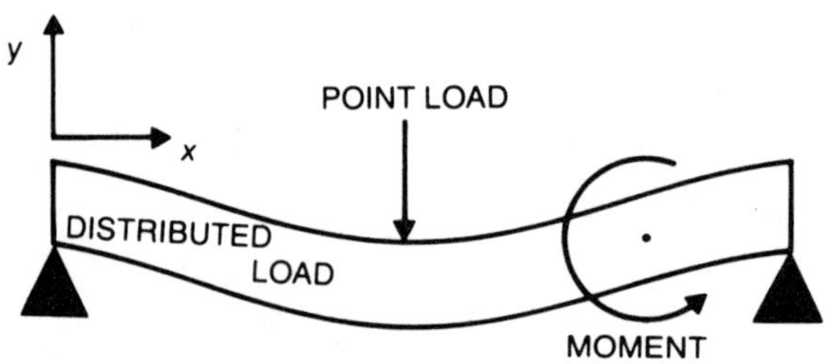

DATA ENTRY

You enter the moment of inertia for the beam, the modulus of elasticity for the beam material, and the beam length. Next enter the point load weight, and its x coordinate. Then you enter the distributed load data, the applied moment, and its x coordinate. The spreadsheet table supplies the results, including the total value of each item. Note that shear is presented in column E. Also note that intermediate calculations (not shown in the spreadsheet) are performed in column A from A50 through A82.

EXAMPLE

You wish to analyze a beam having a moment of inertia of 1 in.^{4}, a modulus of elasticity of 30×10^6 psi, a length of 100 in., a point load of 1000 lb placed at 50 in., and a distributed load of 10 lb/in. The applied moment and distance to the applied moment are both given as zero. The point of interest is at $x = 50$ inches. The spreadsheet is shown in Figure 9.13. Instructions for the spreadsheet are given in Figure 9.14.

	A	B	C	D	E
1		SIMPLE	BEAM		
2					
3		ENTER PARAMETERS:			
4					
5		MO INERT	1	IN^4	
6		ELAST	30000000	PSI	
7		LENGTH	100	IN	
8		PT LOAD	1000	LBS	
9		DIS TO P	50	IN	
10		DIST LD	10	LBS/IN	
11		APP MOM	0	IN-LB	
12		DIS TO M	0	IN	
13		X	50	IN	
14					
15	*********	*********	*********	*********	*********
16					
17		MOMENT	DEFLECT	SLOPE	SHEAR
18					
19	PT LOAD	25000	-.694444	-1.4E-11	500
20					
21	DIST LOAD	12500	-.434028	-6.9E-12	0
22					
23	APP MOM	0	0	0	0
24					
25	TOT	37500	-1.12847	-2.1E-11	500

Figure 9.13: *Simple Beam Spreadsheet*

```
B1:SIMPLE          C1:BEAM
B3:ENTER PAR       C3:AMETERS:
B5:MO INERT        D5:" IN^4
B6:ELAST           D6:" PSI
B7:LENGTH          D7:" IN
B8:PT LOAD         D8:" LBS
B9:DIS TO P        D9:" IN
B10:DIST LD        D10:" LBS/IN
B11:APP MOM        D11:" IN-LB
B12:DIS TO M       D12:" IN
B13:"    X         D13:" IN
A15:/ - *
/R:B15.E15
```

Figure 9.14: *Spreadsheet Instructions: Simple Beam*

```
A19:PT LOAD
A21:DIST LOAD
A23:APP MOM
A25:TOT
B17:"  MOMENT
   B19:@IF(C13>C9,A62,A52)
   B21:-C10*C13*(C13-C7)/2
   B23:@IF(C13>C12,A82,A72)
   B25:+B19+B21+B23
C17:"  DEFLECT
   C19:@IF(C13>C9,A60,A50)
   C21:-C10*C13*((C7^3)+((C13^2)*(C13-(2*C7))))/(24*C5*C6)
   C23:@IF(C13>C12,A80,A70)
   C25:+C19+C21+C23
D17:"  SLOPE
   D19:@IF(C13>C9,A61,A51)
   D21:-C10*((C7^3)+((C13^2)*((4*C13)-(6*C7))))/(24*C6*C5)
   D23:@IF(C13>C12,A81,A71)
   D25:+D19+D21+D23
E17:"  SHEAR
   E19:@IF(C13>C9,A63,A53)
   E21:+C10*((.5*C7)-C13)
   E23:+C11/C7
   E25:+E19+E21+E23
   A50:+C8*(C7-C9)*C13*((C13^2)+((C7-C9)^2)-(C7^2))/(6*C6*C5*C7)
   A51:+C8*(C7-C9)*((3*(C13^2))+((C7-C9)^2)-(C7^2))/(6*C6*C5*C7)
   A52:+C8*(C7-C9)*C13/C7
   A53:+C8*(C7-C9)/C7
   A60:-C8*C9*(C13-C7)*(((C13-C7)^2)+(C9^2)-(C7^2))/(6*C6*C5*C7)
   A61:-C8*C9*((3*((C13-C7)^2))+(C9^2)-(C7^2))/(6*C6*C5*C7)
   A62:-C8*C9*(C13-C7)/C7
   A63:-C8*C9/C7
   A70:-C11*C13*(C12-(C13*C13/(6*C7))-(C7/3)-((C12*C12/(2*C7)))/
   (C6*C5)
   A71:-C11*(C12-(C13*C13/(2*C7))-(C7/3)-(C12*C12/(2*C7))/(C6*C5)
   A72:+C11*C13/C7
```

Figure 9.14, Continued

```
A80: -C11*(C13-C7)*((C7-C12)-(((C13-C7)^2)/(6*C7))-(C7/3)-
(((C7-C12)^2/(2*C7)))/(C6*C5)
A81: -C11*((C7-C12)-(((C13-C7)^2)/(2*C7))-(C7/3)-(((C7-C12)^2)/(2*C7))/
(C6*C5)
A82: +C11*(C13-C7)/C7
```

Figure 9.14: Continued

WOOD COLUMN

PURPOSE

Given the least dimension of a wood column cross-section, the cross-sectional area, and the column length, we use this spreadsheet to determine the slenderness ratio, the critical slenderness ratio, and the allowable column load. Four different types of structural wood are analyzed: Douglas fir, yellow pine, hemlock and southern cypress. The values used for allowable stress and modulus of elasticity for the different woods are those established for select-grade structural timber. The basic equations are:

$$\text{Critical slenderness ratio} = K = 0.64\sqrt{\frac{E}{\sigma}}$$

For $L/D < K$:

$$L = \sigma A\left[\frac{1}{3}\left(\frac{L}{KD}\right)^4\right]$$

For $L/\mathrm{D} > K$:

$$L = 0.4096 \times 2\,AE/3(L/D)^2$$

where:

L = beam length (in.),
D = least dimension (in.),
A = area (sq in.),
E = modulus of elasticity (psi), and
σ = allowable stress (psi).

DATA ENTRY

We enter the value of the least dimension, the cross-sectional area, and the column length. The spreadsheet table calculates the slenderness ratio (*L*/*D*), the critical slenderness ratio (*K*), and the allowable load in pounds. Note that columns E and F are used and that intermediate calculations (not shown) are performed in A24 through D29.

EXAMPLE

We have a 2 × 4 we wish to use as a column of length 8 feet. The least dimension is 1.5 in., the area is 5.25 sq in., and the length is 96 in. Results for each wood type are analyzed in Figure 9.15. Spreadsheet instructions are given in Figure 9.16.

```
          A         B         C         D         E         F
 1                WOOD      COLUMN
 2
 3                ENTER PARAMETERS:
 4
 5                LEAST D        1.5 IN
 6                AREA A        5.25 SQ IN
 7                LENGTH          96 IN
 8
 9  ******************************************************
10
11                L/D=            64
12
13                DOUG FIR YEL PINE  HEMLOCK S CYPRESS
14
15 AL STR          1200      1450       700      1100
16 MOD ELAS     1600000   1600000   1100000   1200000
17 ---------------------------------------------------
18 K=           23.3695  21.25964   25.3704   21.1385
19
20 AL LOAD=         560       560       385       420
```

Figure 9.15: *Wood Column Spreadsheet*

B1:WOOD	C1:COLUMN
B3:ENTER PAR	C3:AMETERS:
B5:LEAST D	D5:" IN
B6:AREA A	D6:" SQ IN
B7:LENGTH	D7:" IN

Figure 9.16: *Spreadsheet Instructions: Wood Column*

```
A9:/ - *
/R:B9.F9
B11:L/D =
   C11: + A22
A15:AL STR
A16:MOD ELAS
B13:DOUG FIR
   B15:1200
   B16:1600000
C13:YEL PINE
   C15:1450
   C16:1600000
D13:HEMLOCK
   D15:700
   D16:1100000
E13:S CYPRESS
   E15:1100
   E16:1200000
A17:/--
/R:B17.E17
A18:K =
   B18: + A23
   /R:C18.E18:R
A20:AL LOAD =
   B20:/FI
   /R:C20.E20
   B20:@IF(A22<A23,A25,A27)
   /R:C20.E20:NRRR
A22: + C7/C5
A23:.64 * @SQRT(B16/B15)
/R:B23.E23:RR
A25: + B15 * C6 * (1 - ((1/3) * ((A22/A23) ^4)))
/R:B25.E25:RNNR
A27:.4096 * (2/3) * C6 * B16/(A22 * A22)
/R:B27.E27:NRNN
```

Figure 9.16, Continued

Appendix: VisiCalc Convention and Command Summary

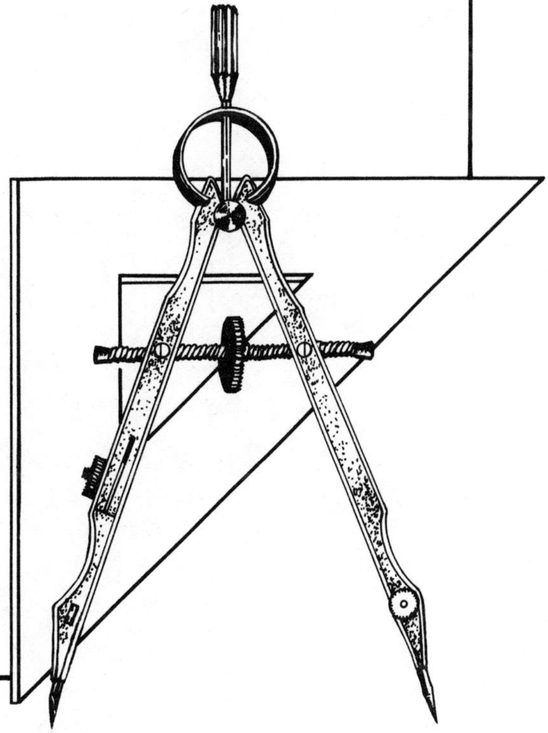

THE VISICALC CONVENTIONS

LABELS

Labels are any combination of letters or symbols used to identify items in a row or column. Labels must always begin with a letter or quote sign ("). You enter a label by moving the cursor to the desired coordinate and typing the label.

Legal label entries

June
Rate %
Jan. 1, 1981
Jan. 1984

Illegal label entries

3 June
.sum
(YTD)
dozen

In general, you can make an illegal label legal by simply placing a quotation sign before the label entry. For example:

(YTD) (illegal)
"(YTD) (legal)

VALUES

Values are numbers—either entered or computed. A value can contain a decimal point, and either a positive or negative sign, but it cannot contain a comma. Values can be entered in scientific notation.

Legal value entries

123
−123.2
1.23E2 (*scientific notation*)

Illegal value entries

$123
123,456

EXPRESSIONS

An *expression* (or formula) can consist of values (numbers), coordinates of other values or expressions, functions, operators, and parentheses. An expression must start with either a digit, a + or − sign, a decimal, a parenthesis, or, if it is the start of a function, an @ sign.

If you use the coordinate of another value in an expression, the VisiCalc program will use the value in the referenced coordinate when evaluating the expression.

Legal expressions

4+3
5*(3+2)
3*@PI
6+B4 (*adds 6 to value in B4*)

Illegal expressions

B4+6 (*needs to start with* +)
3+A (*A is not a coordinate*)

OPERATORS

Legal operators in VisiCalc are +, −, *, /, and ^. These symbols represent:

+ Addition
− Subtraction
* Multiplication
/ Division
^ Exponentiation

The exponentiation operator raises a value to the power indicated.

Examples:

4^2 = 16 (*4 squared*)
4^(1/2) = 2 (*square root of 4*)

Logical Operators

Logical operators compare two values. The logical value True is returned if the comparison is as stated; otherwise, False is returned.

Operator	*Operation*
<	Less Than
>	Greater Than
=	Equal To
<=	Less Than or Equal To
>=	Greater Than or Equal To
<>	Not Equal To

Order of Expression Evaluation

Expressions are evaluated on a strict left-to-right basis. There is one exception, however: quantities within parentheses are always evaluated first. Care must be taken to observe this order when writing expressions; otherwise, incorrect results may occur.

Examples:

```
    4+2= 6
    2+4= 6
  2+4*2= 12
2+(4*2)= 10
 9^(1/2)= 3
   9^1/2= 4.5
```

BUILT-IN FUNCTIONS

VisiCalc contains many *built-in functions* to simplify the writing of frequently encountered expressions. Functions always begin with an @ sign followed by the function itself, and an *argument* in parentheses. This argument may be simply a value (a number or a coordinate position) or an expression (a combination of values and operators).

Examples:

valid functions:

@ABS (−3)
@LN(B5)
@SQRT(B5 + C5)

invalid functions:

ABS(−3)
@ABS−3

Some functions refer to the argument as a list. A *list* may be a combination of values, expressions, and ranges, separated by commas.

A *range* refers to a sequence of column or row positions that begin and end with the stated coordinates. The coordinates are separated by a period.

Examples:

E6.F6
A7.D7

A summary of the VisiCalc functions follows.

SUMMARY OF THE VISICALC FUNCTIONS

@ABS(V) — returns the absolute value of any argument V.

Examples:

@ABS(3) = 3
@ABS(−3) = 3
B5: (−5)
@ABS(B5) = 5

@AVE(list) — calculates the average value of the entries in a list. Blank entries in a list are ignored.

Examples:

@AVE(1,2,3) = 2
@AVE(B1.B3) = average of values in B1, B2, B3

@CHOOSE(V,list) — chooses one of a list of values. The number, V, is the position in the list that is selected.

Examples:

@CHOOSE(3,10,20,30,40) = 30
@CHOOSE(2,B1.B4) = value of B2

@COUNT(list) — results in the number of non-zero entries in a list.

@EXP(V) — gives the result of raising the constant *e* (2.71823) to the V power. This expression is entirely equivalent to (2.71823)^(V).

@LN(V) — computes the natural logarithm of a number, V.

@LOG10(V) — computes the logarithm to the base 10 of a number.

@LOOKUP(V,range) — looks up values in a table. The table contains a range of search parameters and corresponding values. The @LOOKUP function obtains the value from the table corresponding to the next lower search parameter.

@MAX(list) — provides the maximum value in a list.

@MIN(list) — provides the minimum value in a list.

@NPV(i,range) — calculates the net present value of a series of cash flows. The cash flows are to be in the range. The discount rate used is the value i. Cash flows must be in the order to be received and are at the end of the period.

@PI — provides the constant 3.1415926536. Unlike other functions, does not take an argument.

@SQRT(V) — provides the square root of a value and is equivalent to V^(1/2).

@SUM(list) — adds the numbers in a list.

Examples:

B1:2 B2:4 B3:6
@SUM(12,B1.B3,10)=34

TRIGONOMETRIC FUNCTIONS

VisiCalc can compute the trigonometric functions in radians. Legal trigonometric functions are:

@SIN(V)
@COS(V)
@TAN(V)
@ASIN(V)
@ACOS(V)
@ATAN(V)

The last three functions are arc functions.

THE VISICALC COMMANDS

THE **BLANK** COMMAND

Syntax: /B

The BLANK command erases the label, value, or expression at the current cursor location. It does not affect format specification.

THE **CLEAR** COMMAND

Syntax: /C

The CLEAR command erases all information on the current spreadsheet. Before clearing, however, the prompt line asks you to type Y to confirm. CLEAR has the following effects. It:

1. erases all the information
2. places the cursor at A1
3. causes the direction indicator to indicate horizontal on Apple II computers
4. resets formats, titles, and windows to the values that they held when the spreadsheet was loaded.

THE **DELETE** COMMAND

Syntax: /D(R,C)

The DELETE command automatically removes an entire row or column from the VisiCalc spreadsheet. Prior to typing the command, you must place the cursor in the row or column you want to delete. When the command is typed, the prompt line then reads:

DELETE:RC

R deletes the row the cursor is in, and C deletes the column. The VisiCalc program then fills in the deleted row or column as follows:

- The rows below the deleted row all move up one row. The columns to the right of the deleted column move left one column.
- Expressions containing coordinate references are adjusted according to the moves that are made.

Example:

A1: +C1 +2

If column B is deleted, column C will shift left to become the new column B, and then the expression is:

A1: +B1 +2

THE **EDIT** COMMAND

Syntax: /E(ESC,←,→,char)

The EDIT command allows correction of a label, value, or expression, without having to retype it. To use the EDIT command, you place the cursor over the coordinate that is to be modified, and then type /E. The edit cursor is moved left or right with the left- and right-arrow keys. Typing any character inserts that character to the left of the edit cursor. Typing ESC removes the character to the left of the edit cursor.

Example:

A1:4 +3 * 2

To add parentheses to the above expression, first type /E. By pressing the right-arrow key twice, you can then place the edit cursor over the 3. Next, simply type a left parentheses and it will be added to the formula. Finally, press the right arrow four times and type a right parenthesis. The edit line should now read:

A1:4 +(3 * 2)

THE **FORMAT** COMMANDS

Syntax: /F(D,G,I,L,R,$, *)

The FORMAT command is used to modify the information displayed on the VisiCalc screen or hard copy report. The actual stored data is unaffected by this command.

The FORMAT command will modify the format of any coordinate on which the cursor is positioned. You can re-format an entire row by formatting the first position, and then replicating the format across the row. *Caution:* you must do this prior to data entry, or the data in the first positions will be replicated over the entire row. However, an individual position may be formatted without affecting the stored contents.

Let's examine some of the individual format options.

Default (/FD) resets the entry position format to the Global format currently in use. If you have not previously set the Global format, the format will be set to the General format.

General (/FG) displays all values in decimal or scientific notation, depending on the number of significant digits. Labels are left adjusted; values are right adjusted. A leading blank is added to values.

Integer (/FI) displays all values in integer form, rounded to the nearest whole number.

Left Justification (/FL) left-justifies numbers, with one leading blank. Integer format will no longer be in effect in that entry position.

Right Justification (/FR) moves a label to the right, provided the label has fewer characters than the column width.

Dollar Format (/F$) displays all values in dollars and cents format (two decimal places).

Bar Graph Format (/F*) rounds the local value to the nearest integer, and displays the number of asterisks equal to the local integer value.

THE **GLOBAL** COMMAND

Syntax: /G(C(#), O(R,C), R(A,M), F(D,G,I,L,R,$,*))

The GLOBAL command has four options that cause universal action on the entire VisiCalc sheet. If more than one window is in effect, the global refers to the window on which the cursor resides.

Global Column (/GC#) changes the width of all columns to the number (#) entered. Column width may be between 3 and 37. There is no provision for adjustment of individual column widths.

Global Order (/GO(R,C)) determines the order of recalculation for the entire VisiCalc sheet. When the VisiCalc program is first loaded, recalculation order is by columns. That is, first the expression in A1 is evaluated, then A2, etc. After the end of column A, expression evaluation goes to B1. You can change the recalculation order to rows by typing:

/GOR

and then back to columns again, by typing:

/GOC

Recalculation by rows means the order will be A1,B1, . . . A2,B2, etc.

To obtain reliable results, an expression must be evaluated before it is referenced by another expression. If it is not, an error condition called a *forward reference* will occur. You should carefully plan your VisiCalc forms, and carefully choose your order of recalculation, so that foward references do not occur.

Global Recalculation (/GR(A,M)) Each time a value or expression is changed, the VisiCalc program re-evaluates the entire spreadsheet. This can cause undesirable delays. These delays may be overcome by typing:

/GRM

Recalculation will then cease until you type

/GRA

to resume the automatic mode. You may also recalculate by typing an exclamation mark.

Global Format (/GF(D,G,I,L,R,$,*)) works like the FORMAT command we just described, except that it applies to the entire VisiCalc spreadsheet, rather than an individual entry position.

THE **GO TO** COMMAND

Syntax: >COORD

The GO TO command moves the cursor from its current location to the coordinate specified after the "greater-than" sign (>).

THE **INSERT** COMMAND

Syntax: /I(R,C)

The INSERT command inserts a blank row or column into the current VisiCalc spreadsheet. If the INSERT ROW command is used, the row is inserted into the position where the cursor currently resides. The current row, and all subsequent rows, are then shifted down by one row. If the INSERT COLUMN option is used, a column is inserted at the current cursor location. The current row and all subsequent rows are then shifted right by one column.

When the INSERT command is used, expressions containing coordinates are adjusted, so that they reflect the new coordinates.

Example:

A1: +C1+2

If the cursor is now placed over column B and a column is inserted, the new formula for A1 becomes:

A1: +D1+2

Let's now make some additional observations. Many built-in functions include a range of values, for example:

@SUM (B1 . . . H1)

If a column is inserted between B1 and H1, the new sum will be correct and will be:

@SUM (B1 . . . I1)

You cannot insert the column at row B; nor can you insert a column or row in the column or row that contains the sum.

THE **MOVE** COMMAND

Syntax: /M . . . COORD

The MOVE command moves an entire column or row. Expressions are then recalculated to reflect the new coordinates. To move a row down, place the cursor on the desired row, and type /M. Next, type the coordinate in the same column of the row below the destination row.

To move a row up, place the cursor on the desired row, and type /M. Now type the coordinate (in the same column) of the destination row.

To move a column right, place the cursor on the desired column, and type /M. Type the coordinate of the column one column to the right of the desired column.

To move a column left, place the cursor on the desired column. Type the coordinate of the desired column.

THE **PRINT** COMMAND

Syntax /P(P,F)

There are some differences in the use of this command between various computers and printers. Consult your VisiCalc manual for specifics. A general discussion of the PRINT options follows.

PRINT Options

To print your VisiCalc spreadsheets, first, define the portion of the spreadsheet to be printed, by placing the cursor at the upper left coordinate of the portion of the report to be printed, then type /P. A prompt line will appear, asking for the setup string and the lower right coordinate. Consult your VisiCalc manual for the correct response to the setup string. (Often you must type a string or minus sign, in order to obtain the correct number of linefeeds with your computer and printer.) Finally, type in the lower right coordinate that is to be printed.

PRINT FORMAT File

This option of the PRINT command works in exactly the same way as the option described previously, except that the output becomes a diskette file, and is, therefore, not transferred to the printer. This allows the file to be read by a BASIC, Pascal or word processing program.

To exercise this command, you position the cursor at the upper left coordinate of the material you wish to transfer to disk, and then type

/PF (file name)

After typing the file name, respond to the prompt line by typing the lower right coordinate. The PRINT file will be saved on diskette as a standard system file. It may be edited with a text editor, or read by a BASIC or Pascal program, or a word processor.

THE **PRINT FORMULAS** COMMAND

Syntax: /SS,S

The PRINT FORMULAS command prints all formulas and formats for the sheet on the printer. Formulas are printed from the lower right to the upper left coordinate.

THE **REPEATING LABEL** COMMAND

Syntax: /–

This VisiCalc command fills the cursor coordinate position with the sequence of characters indicated after the hyphen. Most often the command is used to draw a line of dashes, equal signs, or asterisks.

As an example, you can draw a line of asterisks across an entire spreadsheet by using the REPEATING LABEL command in combination with the REPLICATE command. This is done as follows:

/–*
/R:A3.H3

THE **REPLICATE** COMMAND

Syntax: /R

The REPLICATE command copies labels, values, and expressions from one coordinate to another. The REPLICATE command can also copy all of the expressions and labels in one row to another row.

Copying Labels and Values To copy labels and values, place the cursor on the coordinate to be copied and invoke the REPLICATE command by typing /R. The program will prompt with:

REPLICATE: SOURCE RANGE OR RETURN

If only a single position is to be copied, you press the return key. If a row or column is to be copied, you must type a period (.) followed by the ending coordinate. The prompt line now becomes:

REPLICATE: TARGET RANGE

You now enter either a single coordinate or a range of coordinates.

Examples:

A1:	
/R:C1:	*(Copy a single entry to a single position.)*
/R:C1.E1:	*(Copy a single entry to multiple positions.)*
/R.A3:C1.E1:	*(Copy a column into three rows.)*

Copying Expressions To copy expressions, you follow the steps discussed above, but, in addition, you must indicate either *relative* or *no* *change* for each coordinate position in the expression. You type N for no change, and R for relative.

As an example, let's assume a constant is located in coordinate C1 and you wish B1 to equal A1 multiplied by C1; B2 to equal A2 multiplied by C1, etc. The sequence below accomplishes this multiplication.

Example:

B1:+C1*A1
/R:B2.B10:NR

(*Note:* In this book, for reasons of legibility, we have put spaces between the groups of N's and R's when there are long replicate sequences. The spaces do not need to be keyed in, however.)

Example:

/R:B3.B10:RRR NRR RRN

THE **STORAGE** COMMAND

Syntax: /S (L,S,D,Q,#)

The STORAGE command loads and saves forms on diskette. The forms stored on diskettes are called files and are referred to by their file names. It is recommended that a file name consist of a recognizable name, followed by a period and an extension. Common extensions are VC for forms, DIF for DIF files, and PF for PRINT FORMAT files. File names can be up to three characters long and must start with a letter.

Examples:

FORECAST.VC (*a report form*)
SALES.DIF (*daily sales in DIF*)
INVENT.PF (*inventory in print format*)

Loading From Diskette (/SL) To load a previously stored file from diskette, simply type /SL file name, where file name is the complete name of the previously saved file.

The LOAD command does not clear any previous forms from memory. If you are loading an individual form, first clear the memory by using the CLEAR command. Although it is possible to combine information from two forms without using the CLEAR command, the two reports must not use overlapping coordinate positions, since only the information from the last loaded form will remain.

The scrolling option of the SAVE command allows you to display one file name at a time. You invoke the scroll by typing a return instead of a file name. Pressing the right arrow key will scroll additional file names.

Saving to Diskette (/SS) You can save files to diskette by typing:

/SS file name

If the file name you select refers to a file already on the disk, the following prompt will appear:

FILE EXISTS: TYPE Y TO DELETE

This will have the effect of replacing the old file with the new.

As previously discussed, you can also use the scrolling option with the SAVE command. You can use the file name selected with the scroll as is, thereby resulting in a file replacement, or you can change the file name, by editing with the escape key. This will create a new file. As shown below, this provides a convenient way to form a series of related files.

Example:

Type /SS followed by a return. The first diskette file title will appear on the display. A series of right arrows will display additional file names. Let's assume that the last file is named SALES3.VC and that you wish to create a file named SALES4.VC.

To create SALES4.VC, you use the normal editing capabilities to change the 3 to a 4. Then you press return. This will add SALES4.VC to the diskette.

Deleting Files (/SD) deletes a named file from a diskette. You may either type in the file name or select it by using the scrolling option.

Quitting VisiCalc (/SQ) provides a way of returning to the computer's main operating system. To return to the operating system, place an operating system diskette in the disk drive and type /SQ. Memory will be cleared and the operating system will be loaded.

Saving in DIF Format (/S#S) The standard *data interchange format* (DIF) provides a way of exchanging data between different VC spreadsheets, and between other programs that use the DIF.

You can specify and save a rectangular area of a spreadsheet. Data and labels are saved in full precision; formulas and expressions are not saved. To use this command, you:

1. Place the cursor at the upper left area of the rectangle to be saved.
2. Type /S#S file name.
3. Type the lower right coordinate.
4. Type R or C, to specify whether the rectangle is saved in row or column order.

Loading (/S#L) loads files already in the DIF. It is not necessary to load files in the same rectangular area or with the same row or column orientation. Thus, a row saved from one spreadsheet may be loaded as a column onto another spreadsheet. To load a DIF file, you:

1. Position cursor at the upper left of the rectangle to be loaded.
2. Type /S#L file name.
3. Type R or C, to specify row or column order.

THE **TITLES** COMMAND

Syntax: /T (H,V,B,N)

To view a report on the computer screen, VisiCalc "scrolls" the data as you move the cursor from one column or row to another. The TITLES command determines that a portion of the screen remains fixed in place; this allows you to continuously view the labels, as the data scrolls.

/TH—causes all the rows at and above the position of the cursor to remain in place. For column labels to remain at the top of the screen, you must first scroll the title row to the top of the screen, by using the cursor controls.

/TV—causes all columns at and to the left of the cursor to remain in place.

/TB—combines the effects of the two previous commands. This allows both horizontal and vertical labels to remain fixed.

/TN—returns the screen to normal appearance.

The **WINDOW** Command

Syntax: /W (H,V,1,S,U)

This command splits the screen horizontally or vertically at the cursor position, thus developing two or more windows. You can display the different windows in different global formats. When operating with two windows, you can move the cursor from one window to another by typing a semicolon.

/WH—splits the screen into two horizontal windows, just above the row containing the cursor.

/WV—splits the screen vertically, just to the left of the column containing the cursor

/W1—returns to the normal one-window condition.

/WS—if two windows are selected, they will initially be scrolled independently. By using this command, you can scroll the windows together.

/WU—returns to unsynchronized scrolling.

Bibliography

Beiser, Arthur, *Applied Physics,* New York: McGraw-Hill, 1976.

Elliot, Robert, *Antenna Theory and Design,* Englewood Cliffs, N.J.: Prentice-Hall, 1981.

Hall, Allen, Holowenko, Alfred, and Laughlin, Herman, *Machine Design,* New York: McGraw-Hill, 1961.

Hecht, Eugene, *Optics,* New York: McGraw-Hill, 1975.

Hergert, Douglas, *Mastering VisiCalc®,* Berkeley, Calif.: Sybex, 1983.

Kreith, Frank and Kreider, Jan, *Principles of Solar Engineering,* New York: McGraw-Hill, 1978.

Liao, Samuel, *Microwave Devices and Circuits,* Englewood Cliffs, N.J.: Prentice-Hall, 1980.

Murrill, Paul, *Automatic Control of Processes,* Scranton, Pa.: International Textbook Co., 1967.

Ramo, Simon and Whinnery, John, *Fields and Waves in Modern Radio,* New York: John Wiley and Sons, 1953.

Rapp, Donald, *Solar Energy,* Englewood Cliffs, N.J.: Prentice-Hall, 1981.

Reference Data for Radio Engineers, 6th ed. Indianapolis, Ind.: Howard W. Sams, 1975.

Roark, Raymond and Young, Warren, *Formulas for Stress and Strain,* New York: McGraw-Hill, 1975.

Trost, Stanley R., *Doing Business with VisiCalc®,* Berkeley, Calif.: Sybex, 1982.

Index

The SYBEX Library

GENERAL INTEREST

DON'T (or How to Care for Your Computer)
by Rodnay Zaks 214 pp., 100 illustr., Ref. 0-065
The correct way to handle and care for all elements of a computer system, including what to do when something doesn't work.

YOUR FIRST COMPUTER
by Rodnay Zaks 258 pp., 150 illustr., Ref. 0-045
The most popular introduction to small computers and their peripherals: what they do and how to buy one.

INTERNATIONAL MICROCOMPUTER DICTIONARY
120 pp., Ref. 0-067
All the definitions and acronyms of microcomputer jargon defined in a handy pocket-size edition. Includes translations of the most popular terms into ten languages.

FOR YOUR IBM PC

USEFUL BASIC PROGRAMS FOR THE IBM® PC
by Stanley R. Trost 174 pp., illustr., Ref. 0-111
A collection of programs for making financial calculations, analyzing investments, keeping records, and many more home, office and school applications.

IBM® PC DOS HANDBOOK
by Richard King 144 pp., illustr., Ref. 0-103
Explains the PC disk operating system, giving the user better control over the system. Get the most out of your PC by adapting its capabilities to your specific needs.

FOR YOUR TRS-80

YOUR COLOR COMPUTER
by Doug Mosher 350 pp., illustr., Ref. 0-097
Patience and humor guide the reader through purchasing, setting up, programming, and using the Radio Shack TRS-80/TDP Series 100 Color Computer. A complete introduction to the color computer.

THE FOOLPROOF GUIDE TO SCRIPSIT™ WORD PROCESSING
by Jeff Berner 225 pp., illustr., Ref. 0-098
Everything you need to know about SCRIPSIT—from starting out, to mastering document editing. This user-friendly guide is written in plain English, with a touch of wit.

FOR YOUR APPLE

THE APPLE® CONNECTION
by James W. Coffron 264 pp., 120 illustr., Ref. 0-085
Teaches elementary interfacing and BASIC programming of the Apple for connection to external devices and household appliances.

BUSINESS & PROFESSIONAL

INTRODUCTION TO WORD PROCESSING
by Hal Glatzer 205 pp., 140 illustr., Ref. 0-076
Explains in plain language what a word processor can do, how it improves productivity, how to use a word processor and how to buy one wisely.

INTRODUCTION TO WORDSTAR™
by Arthur Naiman 202 pp., 30 illustr., Ref. 0-077
Makes it easy to learn how to use WordStar, a powerful word processing program for personal computers.

PRACTICAL WORDSTAR™ USES
by Julie Anne Arca 200 pp., illustr., Ref. 0-107
Special applications for essential office tasks are explained in step-by-step detail. Makes using WordStar efficient and fun.

MASTERING VISICALC®
by Douglas Hergert 217 pp., 140 illustr., Ref. 0-090
Explains how to use the VisiCalc "electronic spreadsheet" functions and provides examples of each. Makes using this powerful program simple.

DOING BUSINESS WITH VISICALC®
by Stanley R. Trost 260 pp., Ref. 0-086
Presents accounting and management planning applications—from financial statements to master budgets; from pricing models to investment strategies.

BASIC

FIFTY BASIC EXERCISES
by J. P. Lamoitier 232 pp., 90 illustr., Ref. 0-056
Teaches BASIC by actual practice, using graduated exercises drawn from everyday applications. All programs written in Microsoft BASIC.

BASIC EXERCISES FOR THE APPLE®
by J. P. Lamoitier 250 pp., 90 illustr., Ref. 0-084
This book is an Apple version of *Fifty BASIC Exercises.*

BASIC EXERCISES FOR THE IBM® PERSONAL COMPUTER
by J. P. Lamoitier 252 pp., 90 illustr., Ref. 0-088
This book is an IBM version of *Fifty BASIC Exercises.*

BASIC EXERCISES FOR THE ATARI®
by J.P. Lamoitier 251 pp., illustr., Ref. 0-101
This is the ATARI version of *Fifty BASIC Exercises.*

YOUR FIRST BASIC PROGRAM

by Rodnay Zaks 150 pp., illustr., Ref. 0-092

A fully illustrated, easy-to-use, introduction to BASIC programming. Will have the reader programming in a matter of hours.

BASIC PROGRAMS FOR SCIENTISTS AND ENGINEERS

by Alan R. Miller 318 pp., 120 illustr., Ref. 0-073

This second book in the "Programs for Scientists and Engineers" series provides a library of problem-solving programs while developing proficiency in BASIC.

CELESTIAL BASIC: Astronomy on Your Computer

by Eric Burgess 300 pp., 65 illustr., Ref. 0-087

A collection of BASIC programs that rapidly complete the chores of typical astronomical computations. It's like having a planetarium in your own home! Displays apparent movement of stars, planets and meteor showers.

PASCAL

INTRODUCTION TO PASCAL (Including UCSD Pascal™)

by Rodnay Zaks 420 pp., 130 illustr., Ref. 0-066

A step-by-step introduction for anyone wanting to learn the Pascal language. Describes UCSD and Standard Pascals. No technical background is assumed.

THE PASCAL HANDBOOK

by Jacques Tiberghien 486 pp., 270 illustr., Ref. 0-053

A dictionary of the Pascal language, defining every reserved word, operator, procedure and function found in all major versions of Pascal.

INTRODUCTION TO THE UCSD p-SYSTEM™

by Charles W. Grant and Jon Butah 300 pp., 10 illustr., Ref. 0-061

A simple, clear introduction to the UCSD Pascal Operating System; for beginners through experienced programmers.

PASCAL PROGRAMS FOR SCIENTISTS AND ENGINEERS

by Alan R. Miller 374 pp., 120 illustr., Ref. 0-058

A comprehensive collection of frequently used algorithms for scientific and technical applications, programmed in Pascal. Includes such programs as curve-fitting, integrals and statistical techniques.

OTHER LANGUAGES

FORTRAN PROGRAMS FOR SCIENTISTS AND ENGINEERS

by Alan R. Miller 280 pp., 120 illustr., Ref. 0-082

Third in the "Programs for Scientists and Engineers" series. Specific scientific and engineering application programs written in FORTRAN.